Rainforests
Land use options for Amazonia

Craig Johnson,
Education Officer, WWF United Kingdom

Richard Knowles,
Head of Humanities, Ogmore Comprehensive School, Bridgend, Mid-Glamorgan

Marcus Colchester,
Projects Director, Survival International (Yanomami Indian chapter)

OXFORD UNIVERSITY PRESS • WWF United Kingdom
in association with
Survival International

**In Memory of
Chico Mendes**

Co-published in 1989 by
Oxford University Press and WWF United Kingdom

Copyright WWF
ISBN 0 19 913339 5

All rights reserved. No part of this publication may be reproduced,
stored in a retrieval system, or transmitted, in any form or by any means
electronic, mechanical, photocopying, recording or otherwise, without
the prior permission of the publisher. Exception: for educational purposes
the 7 photographs on pages 54 and 55 may be photocopied.

Maps by: Ann Wilson

Front cover photos: Yanomami Indian *(Victor Englebert/Survival International)*
　　　　　　　　　Forest scene *(Heather Angel)*

75% of royalties to the work of WWF and Survival International

Printed in Great Britain by Martin's of Berwick

The world's rainforests

'When we say we're against deforestation people say we're against the development of Brazil. We're not against development, but we are against the devestation of Amazonia. We want development that doesn't only benefit the big companies and the powerful, but the people that work on the land.'

CHICO MENDES,
1st National Meeting of Rubber Tappers,
Brasilia, 1985.

'The integration I acknowledge must be based on a profound respect for the community characteristics of each nation. It must preserve the Indian, not make him into one more picturesque figure for the annals of history or the cowboy films of the future. Their technicians are there, discussing our future and what is best for us, without taking into account, even as a polite gesture, what we think and feel in relation to all this. The Brazilian Indian cannot wait silently for plans and studies to be made behind desks and in offices, while progress reaches us. And how does it reach us? In the worst of ways, its diseases, its vices, its degeneration.'

MARIO MARCOS SERENA,
Brazilian Indian leader.

'I am not against gold prospectors, I am against gold prospecting because it makes holes and ruins the river and the river channels. The Yanomami do not do that, cut the ground, cut the trees, burn the forest. We are not enemies of nature. We are friends of nature because we live out there in the jungle. It takes care of our health. Omami (our ancestral hero) gave us the land to live on, not to sell. Whites sell and then go to another place. Indians do not do that ... Now the fish are suffering, the rivers are being destroyed. Even the whites are suffering up there. Indian and White, poor White and rich White. Because the sickness is not afraid, it kills anybody – the rich, the brave, the big ones. My land is the last land to be invaded, it is the last invasion. After the Indian suffers, the White will suffer too. And then the war will come among you ...'

DAVI KOBENAWA YANOMAMI,
A leader from a Yanomami community on the Demini River, on receiving his United Nations Environment Programme Global 500 award,
January, 1989.

CONTENTS

CHAPTER ONE   **INTRODUCTION – ECOLOGY OF THE AMAZONIAN RAINFOREST**

Page

An introduction by Dr Norman Myers   9

CHAPTER TWO   **THE TRADITIONAL ECONOMY**

2.1	Traditional land uses in Amazonia: an introduction	13
2.2	Location and habitat of the Yanomami Indians: the Sierra Parima	14
2.3	The Yanomami today	15
2.4	Traditional Yanomami patterns of land use	18
2.5	Brief comparison with other Amazonian groups	24
2.6	Amazonian land use practices in transition	25
2.7	The impact of imposed development	25

CHAPTER THREE   **THE EXTENT OF DEFORESTATION – AMAZONIA AND BEYOND**

3.1	Monitoring deforestation in Amazonia	28
3.2	Deforestation pattern at the regional level in Amazonia	29
3.3	Likely future patterns of deforestation in Amazonia	29

RAINFOREST DESTRUCTION AT THE GLOBAL LEVEL

3.4	Monitoring global deforestation – the UN Tropical Forest Assessment Project	30
3.5	The UN tropical forest classification system	31
3.6	UN deforestation estimates in tropical forests	34

CHAPTER FOUR   **THE IMPACT OF NEW AGRICULTURAL DEVELOPMENT**

4.1	Brief overview of recent land uses in Amazonia – an historical perspective	35
4.2	Recent pioneer smallholder colonisation, cattle ranching and infrastructure development	36

4.3	Factors promoting migration of pioneer smallholders into Amazonia	41
4.4	Pioneer smallholder annual crops	41
4.5	Cattle ranching	44
4.6	Other agriculture on terra firme	45
4.7	Farming on varzea	45
4.8	Human carrying capacity of land, land tenure distribution and human consumption patterns	46

CHAPTER FIVE — THE IMPACT OF REGIONAL PLANS: AGRICULTURE AND INDUSTRY

5.1	Regional plans for Amazonia: an introduction	47
	5.1.1 Energy development – hydro-electric power	47
	5.1.2 Mineral extraction	48
5.2	Polonoroeste and the World Bank	50
	5.2.1 Recent history and description of Polonoroeste	51
	5.2.2 Criticisms of Polonoroeste and World Bank involvement	56
	5.2.3 NGO lobbying of World Bank and US Congress	57
	5.2.4 The World Bank's response to criticisms about Polonoroeste	59

CHAPTER SIX — TOWARDS BALANCED DEVELOPMENT

6.1	The Amazon Pact and the Brazilian Government's Forest Policy Report	61
6.2	Rubber tapper "extractive reserves": a new conservation – development initiative for Amazon forests	63
6.3	Identifying and implementing an "appropriate" Amazonian development strategy	65
	6.3.1 An introduction	65
	6.3.2 Evaluating and ranking development options for Amazonian "terre firme" rainforest	66
	6.3.3 Comparison of development options	67
	6.3.4 Conclusions	72

References — 76

Acronyms — 77

Glossary — 77

Acknowledgements — 78

FIGURES

		Page
Figure 1	Typical Yanomami garden	20
Figure 2	Differences between "forest" and "river" Indians	25
Figure 3	Deforestation in Brazil's Legal Amazon	30
Figure 4	UN tropical forest classification system	32
Figure 5	Global areas of tropical woody vegetation	33
Figure 6	Main road network in Amazonia	39
Figure 7	Main migratory flows during the 19th and 20th centuries	40
Figure 8	Population of the Amazon region, 1950–1980	40
Figure 9	The relationship between deforestation and crop failure	42
Figure 10	Changes in phosphorus after conversion from forest to pasture	44
Figure 11	A varzea utilisation scheme	46
Figure 12	Map showing Tucurui hydro-electric power project and Carajas minerals programme	49
Figure 13	Map of Northwest Region Programme (Polonoroeste)	53

CHAPTER 1

Introduction – Ecology of the Amazonian Rainforest

Dr Norman Myers

Amazonia evokes all manner of images. Of an "endless" sea of vegetation, usually tangled. Of creatures teeming and unknown, fascinating at a distance but often fearful at close quarters. Of landscapes laden with moisture, and daily downpours as the norm. Of a giant climatic flywheel, moderating weather patterns far and wide. And finally, of a last wild region that is to be tamed by pioneering man.

Certainly Amazonia generates a welter of "ooh and ah" facts and figures. It first became known to outsiders when a Spanish sea captain, sailing in the western Atlantic Ocean along the equator, was surprised to find he was surrounded by fresh water. Heading westward to investigate, he found he was actually in the outflow of a giant river. The Amazon drains an area of 6.2 million square kilometres, and flows 6 762 kilometres (as far as from London to New Delhi) before reaching the Atlantic. There it disgorges one-fifth of all riverwater on Earth, or four times more than the second-largest river, the Zaire, and eleven times more than the Mississippi. It delivers as much water in one day as the Thames in one year. For a fair distance upstream, the Amazon is wider than the English Channel, that is, a good 50 kilometres; from a boat in the middle of the river a passenger cannot always see the shore on either side. The mouth of the river is 240 kilometres across, embracing in its delta an island called Marajo, which is the size of Switzerland. The outflow stains the ocean a muddy colour for hundreds of kilometres out from land.

The entire river system of Amazonia includes 17 tributaries that are more than 1 600 kilometres long. Only recently scientists have discovered several new tributaries of the Amazon River. We now know too that we have formerly drawn entire mountain ranges on our maps of Amazonia in places hundreds of kilometres away from their actual location. The collective length of all the Amazon's tributaries, 10 000 of them altogether, is more than 80 000 kilometres, or twice as much as would stretch around the equator. The two largest tributaries, the Negro and the Maderia, are each as large in water volume as the Zaire River. At least 24 000 kilometres of waterways serve as navigable "trunk rivers", representing a magnificent network of communications and undermining Brazil's declared rationale for the Trans-Amazon Highway system as a device to connect the further sectors of the basin.

The vast river system is almost completely flat. After its sources in the Peruvian Andes, less than 190 kilometres from the Pacific shoreline, the river falls almost 5 000 metres in its first 1 000 kilometres, but in the remaining 5 800 kilometres it drops to sea level by only about 300 metres, a scarcely discernible slope. The Brazilian city of Manaus, 1 600 kilometres upstream, is a mere 30 metres above sea level, so the river's drop is about 1.5 centimetres per kilometre. This flatness has a favourable consequence in that it allows ocean-going ships to penetrate far inland – a distance equivalent to crossing the North Atlantic. It also allows the ocean's tidal effects to be strongly felt 600 kilometres upstream from the mouth. Is it any wonder that early explorers spoke of the Amazon as the "rio-mar", or the "ocean river"?

Although the main catchment zones receive rain throughout the year, the melting snows in the Andes increase the volume of water so much that they cause downstream flooding. At Manaus the annual fluctuation in river level averages almost 12 metres, and high waters of more than 18 metres are not unknown. While this phenomenon makes things difficult for the burghers of Manaus, it makes life exceedingly fecund in the extensive floodplains. These **varzeas**, as the floodplains are known, extend at least 50 and sometimes 100 kilometres from the main channel. They cover 100 000 square kilometres, or almost 2 percent, of the basin and support the largest, most complex and most variegated "flood forests" known anywhere.

An indication of the ecological complexities in this strip sector of Amazonia is the large number of fish that feed primarily off seeds that fall from trees. Many of the fish apparently feed off nothing else during low-water months, but draw instead on the fat reserves they have built up in their large stomachs. In order to adapt to their specialized diets, certain of the fish have also developed large flat-topped molars (similar to those found in sheep, cows and other animals that graze), together with jaw musculatures that enable them to crack even tough Brazil nuts. Through their feeding activities, the fish help to prepare nuts for germination. They also distribute fruit seeds, which further assist the trees in return. As many as 200 fish and tree species appear to depend on each other, and of the fish at least 50 are commercial species, making up three-quarters of the fish catch in Amazonia.

These ecologic linkages deserve recognition when development technocrats assert that, were the forest cover to be cleared, the floodplains could grow vast quantities of rice and support huge herds of water buffalo. So they could, but in many localities, better-quality food could be obtained through maintaining the flooded forests in order to harvest the sustainable bounty of the rivers' fish stocks. The Amazonia fishery can be roughly estimated at a minimum of one million tonnes, with a renewable offtake of one-quarter of a million tonnes; or as much animal protein for meal tables as could be produced from 25 000 square kilometres of Amazonian cattle ranches.

As is well known, Amazonia contains many species. The region harbours 2 000 known species of fish, or eight times as many as the Mississippi River system, and ten times as many as in the whole of Europe. The eventual total of fish species to be identified may reach 3 000. Away from the rivers and mostly on the **terra firme** (slightly higher lands) live one in five of the world's birds in scarcely one-fifteenth of the Earth's land surface.

All in all Amazonia contains several million animal species, most of them invertebrates, especially insects. A century ago the British explorer Henry Walter Bates collected more than 700 species of butterfly within just an hour's walk of his home in eastern Amazonia, amounting to about one in 30 of all butterfly species on earth. Today a student scientist could surely find a new insect species within a single day's search at most and have it named after him/her.

Some of these insects reveal the intricacies of Amazonia's ecology. An iridescent insect, a member of the euglossine group of bees, is the sole pollinator of the Brazil nut tree, well known for its tasty nut. The bee also pollinates **aroids**, among other plants, and in turn the **aroids** often supply prime sources of food to sundry other insects, which pollinate further plants, and so on. In this crucial sense, we can view the euglossines as "keystone" species in their ecosystems.

Regrettably we shall never know how truly rich Amazonia is in species. Botanists believe another 10 000 plant species, for instance, wait to be discovered. They are equally sure they will not get to the plants' habitats and identify them in time. At present rates of deforestation, the chainsaw man will beat them to it. The situation would be much improved if there were enough Amazonian botanists to go and get on with the job. But Colombia, a country of 1 179 369 square kilometres and probably an extraordinary 20 000 plant species in its sector of Amazonia, has fewer than 20 botanists with expertise to go out and identify a new species. Britain, by contrast, with its 244 001 square kilometres, has more than 1 500 botanists, and fewer than 1 500 plant species for them to look at.

So we know well enough that Amazonia contains many species. Not so well known is that almost all of these species are very sparse in number. In a typical one hectare patch of Amazonia with its 100 or more

Introduction – Ecology of the Amazonian Rainforest   **11**

Orchid species. *(B. Rogers/Biofotos)*

Orchid species. *(B. Rogers/Biofotos)*

Ocelot. *(F. Vollmer/WWF)*

Arrow poison frog. *(Heather Angel)*

tree species: contrast a mere dozen or so in a British forest patch of similar size. As many as half the species may be different from those in another hectare only two and a half kilometres away; whereas in Britain they are all likely to be identical. Thanks to these remarkable distribution patterns, many of Amazonia's plants are comparatively rare, which makes them all the more liable to local extinction when the forest is cut back. Much the same applies to birds among other animals. The black-chested tyrant flycatcher recently appeared 800 kilometres from the nearest place it had hitherto been seen. While the red-shouldered parrot had been recorded only through sitings of seven isolated individuals in all the 6.2 million square kilometres of Amazonia before scientists recently encountered an entire flock.

As a final instance of Amazonia's remarkable traits, consider its climate, and how it interacts with the forest. The wettest region on Earth, Amazonia contains two-thirds of all fresh water on Earth, in the sense of "free" fresh water above ground; much fresh water is "locked up" in polar ice caps and in subterranean reservoirs. This water is constantly falling from the sky, percolating through the complex layers of vegetation, being absorbed by plants, running away into streams and rivers, and evaporating through the sun's warmth or being evapotranspired by plants back into the atmosphere. Surprisingly enough, more than half of the region's moisture remains within the Amazonian ecosystem. That is, it does not make its way into the ocean before being blown back over the land again. Rather, as soon as the moisture (most of it anyway) falls to the ground, it is returned, through the respiratory activities of trees and other plants, into the skies, whereupon it gathers for a fresh series of thunderstorms. Day in, decade out, the water cycles round and round, remaining within the bounds of Amazonia.

The implications are profound. Were a substantial amount of forest to be removed, the climate would become drier than at present. Each time a sizeable sector of forest is cut down, the remainder is less capable of evapotranspiring as much moisture as was circulating through the ecosystem before. Clearly this makes for a steadily desiccating ecosystem. At what stage could the forest start to be transformed into a different kind of forest by virtue of the drying-out phenomenon? Has the process already begun? If so, how far has it gone? Could it still be reversed? To date, we scarcely know how to ask the correct questions, let alone to supply the right answers.

Equally to the point, we are now learning that the climate of extensive territories in Brazil outside Amazonia depends, in part at least, upon the same hydrological cycles that are so critical to the persistence of the rainforest as we presently know it within Amazonia. South of Amazonia lie vast cerrado woodlands, these being open woodlands with scattered trees, very different tree formations from Amazonia. If Amazonia were to be fundamentally transformed through deforestation, there could well be less rainfall across most of the cerrado expanse. Much the same could even apply to Brazilian territories still farther south and west, including the principal agricultural sectors of the country. Fortunately, the Brazilian government is starting to take note of these climatic linkages, in their full scope so far as they can be discerned at this early stage, and Brazilians are becoming less inclined to look on the Amazonian forest ecosystem as a "valuable asset going to waste".

This, then, is what is at stake in Amazonia: a forest ecosystem that is biologically richer and ecologically more complex than we can fairly apprehend, let alone comprehend. But this applies only insofar as Amazonia remains largely untouched by human hand. In this book you are going to read much about the human hand that is now modifying parts of the region beyond recognition, and may, within your own lifetimes, eliminate virtually all the region's rainforest together with similar rainforests right around the tropics. Were the next few decades to see the end of the greatest celebration of nature to grace the planet during four billion years of life's history, we would need to re-colour our atlases: where there is now a band of rich green around the equator, depicting lush vegetation, we would need to replace it with dirty brown or pale grey, to denote the impoverished vegetation that had replaced the rainforest. Do we really want to impose such a basic and unprecedented change on our atlases, and on our Earth?

CHAPTER 2

The Traditional Economy

2.1 Traditional land uses in Amazonia: an introduction

The Amazon region was largely peaceful throughout the 19th and early 20th century, with only very gradual changes as small numbers of settlers of Portuguese extraction moved into the rainforests, home for a wide diversity of Indian groups. Until this century the main economic activity carried out by the non-Indian settlers was collection of natural products, mainly Brazil nuts, cocoa, cloves, cinnamon, precious woods, turtles and fish from the rivers. The descendants of those early settlers are the "caboclos", who are usually of mixed Indian and Portuguese race. They live in the forest as agriculturalists and as hunter-gatherers. Their small clearings have cassava, some fruit trees and often a few chickens and pigs to complement the foods they take out of the rainforest. Products taken from the forest include tropical fruits, game and fish. The present density of caboclos is low so their communities have minimal impact on the forest ecosystem.

Throughout the 19th century various attempts to set up large-scale commercial agriculture were made, but none of them prospered. One of the problems was shortage of labour, a difficulty which became particularly marked after the abolition of slavery in 1888. It was however, the rubber boom at the turn of this century which finally killed off commercial farming.

Mary Helena Allegretti and Stephan Schwartzman (1986) record that most of the rubber tappers ("seringueiros") who live and work in Amazonia's rainforests today are descendants of the poor northeasterners who migrated to Amazonia between about 1870 and 1910 during the Amazonian rubber boom. Following Goodyear's discovery of the vulcanization process in 1839, and the emergence of mass-produced machines, demand for rubber grew enormously (especially for automobiles). The Brazilian Amazon produced most of the supply. During the boom period, production of rubber for export increased from some 156 tonnes in 1830 to 34 248 in 1910. Within the next decade, production fell drastically, and prices even more so, as a result of the introduction of plantation rubber in southeast Asia. This followed Sir Henry Wickham's clandestine removal of Brazilian rubber seeds to the Kew Botanical Garden in 1876. Migration to Amazonia estimated at some 260 000 people between 1872 and 1900, ceased as the labour market contracted. The fabulous fortunes gained in the boom years at the expense of mass exploitation of rubber tappers and Indians, were spent in ostentatious display or invested in other sectors. Amazonia's contribution to world markets soon returned to insignificance.

With the collapse of the region's rubber economy, reverse migration to the northeast cities occurred, but not all the rubber tappers left Amazonia. Some continued to work for patrons. Others remained in abandoned rubber producing areas, and developed a semi-autonomous mixed economy, based on small-scale agriculture, hunting, fishing, gathering and latex production for sale.

Shifting cultivation, also known elsewhere as "swidden" or "slash and burn", is the traditional method of farming Amazonia's vast unflooded uplands. These uplands are known in Brazil as "terra firme". Indigenous Amerindians have used shifting cultivation for many centuries as a way of obtaining food crops from Amazonia's infertile soils with a minimum of human effort spent on fending off the relentless competition of weeds and crop pests. Shifting cultivation by the Yanomami Indians is studied in this chapter. Yanomami land use illustrates an economy once typical of large areas of Amazonia.

The caboclos also employ shifting cultivation. Caboclos generally do not move their villages together with their fields, as Amerindians often do. However, they are still able to move their plantings over sufficiently wide areas to have long "bush fallows". Most of the region's river and stream banks now occupied by the caboclo population have been farmed by these people for about 100 years. This is in contrast to the much longer history of occupation, very often of the same choice riverside sites, by Indian groups.

2.2 Location and habitat of the Yanomami Indians: the Sierra Parima

The Yanomami inhabit a stretch of country on both sides of the Venezuelan–Brazilian border. Lying between 0 and 6 degrees north and 61 and 67 degrees west, the area lies on the watershed between the Orinoco and Amazon basins. This area, known as the Sierra Parima, forms the south-western part of the Guyana Highlands, an ancient upland region that is part of the core of the South American continent. The highlands form the uplifted part of the Guyana Shield. High mesas, locally known as "tepuis", made of sandstone stand out in a landscape that has been extremely eroded to expose, over the majority of the area the much older basic crystalline shield on which the Cretaceous sandstones lie unconformally.

Apart from the spectacular tepuis some of which reach heights of over 2000 metres, the area is characterised by steep rolling hills rising to over 1000 m on the watersheds. The major part of this area is covered with dense rainforests that receive between 2.5 and 4 metres of rain per year.

Woman with nose sticks. The Yanomami are one of the last great Indian nations of Amazonia who still maintain their traditional life. *(Victor Englebert/Survival International)*

Most of the soils and alluvia in the area derive from these sandstones and the underlying basic shield. The soils are for the most part very poor, containing little in the way of nutrient stores and in many parts being little more than pure sand.

Being right on the equator, the Yanomami area does not experience the exaggerated seasonality that prevails in temperate climates. Mean monthly temperatures scarcely oscillate throughout the year – mean maximum temperatures at 500 metres being about 30°C year round. Despite the uniformly warm and very moist conditions – under the forest canopy humidity approaches saturation for much of the time – there is nevertheless a distinguishable seasonality present. During the local "summer", the "dry" months of September to April, precipitation decreases perceptibly, cloud cover lessens, maximum and minimum temperatures diverge more greatly and evaporation increases. For the rest of the year the rains increase again, rivers are flooded and the larger game animals that concentrated in the moister parts of their ranges become again more dispersed.

Forest view with huts. The Yanomami live in some 360 communities spread out over an area of tropical forest the size of England and Wales. *(Victor Englebert/Survival International)*

The Indians' habitat is thus one of streams and swiftly-falling rivers interrupted by innumerable small rapids and waterfalls, making navigation impracticable; of steep valleys and mountains whose summits sometimes shake clear of the enclosing vegetation to reveal their worn and stony crowns; and, above all, of dense, cool rainforest broken now and then by small clearings that the Yanomami have cut from the forest: in these hot, bright spaces the Yanomami sow their crops and make their simple huts.

2.3 The Yanomami today

Yanomami is the popular term used to refer to the Yanoama, a group of peoples inhabiting the forested uplands at the headwaters of the Orinoco and the northern tributaries of the Amazon. Numbering over 21 000, they are the largest of the few remaining Indian nations in Amazonia that still live essentially independently of the market economies of the two national societies in which they now find themselves.

Relatively isolated from the rest of Amazonia by the impenetrability of their upland region, cut off from easy access with the lower rivers by high waterfalls and numerous rapids, the Yanomami have survived and thrived in their forested homeland while the other Indian peoples have been progressively decimated and destroyed by the shattering impact of western societies.

Until recently the Yanomami, who depend on hunting, fishing, collecting and "slash and burn" agriculture for their subsistence, were expanding physically and numerically. In the 1950s, at a time when the last and most isolated communities finally gave up the use of stone tools, their territories began to be invaded by missionaries and miners, who with the advent of easily available monoplanes and outboard engines, were able to reach areas long closed to non-Indians. Since then, the Yanomami's lands have been under increasing attack, culminating in the early 1970s with a road construction programme, by which the Brazilian government thrust a road across their territories causing a massive loss of life to communities in the contact zone. Especially in the mining areas, but also in the Upper Orinoco, the Yanomami peoples are still suffering severely the repeated epidemics caused by the transmission of diseases previously unknown to them. Mortalities of as high as 30% from one single epidemic have been recorded in some villages, and overall mortalities in some villages near the road have approached 90%. The effects on the Yanomami's social life have been shattering, many of the survivors adopting a roadside existence as beggars and prostitutes.

In the 1980s the assault on the Yanomami's lands has only accelerated, with attempts being made, both in Venezuela and Brazil, to mine gold and tin in the very heart of the Indians' lands. Menaced on all sides, the Yanomami are themselves now coming to realise their serious predicament and have begun to speak out openly against these invaders of their lands. They have joined their voices to those, like the human rights organisation *Survival International*, who have campaigned for a recognition of Indian land rights.

The Yanomami speak out

'There are many Yanomami who live in Brazil and Venezuela. We think that there are more than 20 000. We are all one people because we understand the language of them all.

'The Yanomami do not go out from their villages and so most of them only speak Yanomami. I, Davi, first studied in my own language, I have started to read and write in Yanomami. I have never been in a school for white people and so I can't speak Portuguese well. The other Yanomami also don't speak Portuguese.

'We were invited to this meeting to tell you of our situation. Our lands have not been demarcated. That is

The Traditional Economy   17

why we are being invaded by the whites who are taking gold from our lands and are bringing diseases and contaminating the Yanomami. We call the white man's diseases *xawara*. These diseases kill our people.

'At first we didn't know that the miners were invading our land. Now we know: those who live near the miners and ranchers, the Yanomami of the Ajarani, the Catrimani, the Demini, the Couto do Magalhaes and the Erico. There are Yanomami that know that it is bad for them and are sorry because they are striken with illnesses. There are others who think it is good because they receive machetes, axes, pans and matches that they use in the forest.

'But we, who know that the miners deceive us, are telling the others. So that they know what is happening. The miners want to take our Yanomami women to keep them and they are deceiving us and stealing our gold.

'I am telling you this because I am worried and angry. I want you to know of our situation, to understand our worries and to join our struggle.

'We Yanomami want the demarcation of our Yanomami Park. A continuous area, that is very important for us Yanomami.'

Statement issued by Davi, Yanomami from Toototobi, Carreira, Yanomami headman of Wakathaotheri, and Rubi, Yanomami from the lower Catrimani, at a meeting at Surumu, 7 to 9 January 1985.

2.4 Traditional Yanomami patterns of land use

Social organisation

The Yanomami live in over 360 settlements distributed over an area twice the size of Wales, giving them an overall population density of some 9 km²/person.

The small communities, which average about 50 inhabitants, are widely dispersed from one another and many are separated from their nearest neighbours by a trek of one or two days' length. Communication between settlements is mainly by foot. Only in the peripheral areas, where the rivers are larger and contact with other Indians has occurred, have the Yanomami adopted the use of canoes, skilfully made by hollowing out the trunk of a single tree.

Yanomami villages are generally made up of several extended families linked to each other through many generations by marriages. Ties of marriage and descent also link together neighbouring communities, which frequently come together for festivals. Yet large villages are inherently unstable among the Yanomami.

Yanomami social life places a strong emphasis on the independence, rights and also the obligations of the *individual*. The Yanomami are brought up to be self-reliant, to stand up for themselves and exact revenge for the wrongs done to them, in defence of their own interests. On the other hand, Yanomami are also imbued with a strong sense of the need to share their goods with others. Inequalities of wealth and status are minimal and yet it is hardly an exaggeration to say that the only real "sin" in Yanomami life is selfishness.

An intensely egalitarian people, their society functions without centralised power structures or figures vested with authority over others. There are no "chiefs" in Yanomami society, though kin groups do promote their interests through spokesmen, who act as leaders particularly in times of conflict.

An important consequence of this emphasis on equality and the lack of authorities is that there are no equivalents to courts or a judiciary to resolve disputes. Differences of opinion are easily tolerated, but more serious disagreements are more often solved through the splitting of communities or by one part of the village temporarily moving elsewhere.

Moreover, the Yanomami are a warrior people. Serious conflicts sometimes become violent and may lead to the permanent division of villages and even to raiding between communities. Warfare and feuding are common.

The social order thus functions to keep the communities small, mobile and widely dispersed, a fact that has important implications in spreading the overall pressure exerted on the environment.

Slash and burn agriculture

The Yanomami derive the bulk of their diet from their gardens. The gardens are cut from the forest with axes and machetes (bush knives) which are today obtained mainly from the church missions. The underbush is cleared first. Then the major trees are felled, frequently by partially cutting through the smaller trunks and then felling a forest giant which brings down the smaller trees as it crashes to the ground. The fallen vegetation is then left to dry out for a number of weeks during the dry season.

Once sufficiently dry, the felled vegetation is fired and is burnt. The burn destroys much of the nutrient store of the forest as the flames oxidise and volatilise many of the nutrients locked up in the vegetation mass. Nevertheless the remaining ash still retains an important portion of nutrients which are released into the soil by the rains. Equally important is the fact that the burn kills off most of the seeds and seedlings in the soil, thus preventing the immediate invasion of the plot by weeds.

The heavy rains act fast to wash most of the nutrients derived from the burnt vegetation into the soil. Here they are quickly leached away leaving the soils as poor as ever. In order to take advantage of the ephemeral abundance of soil nutrients, the Yanomami therefore commence planting immediately with the onset of the rains. The Yanomami plant the suckers of bananas and plantains. They also plant the cut stems of **cassava**, corms of cocoyams and small sweet potato and yam tubers and sow other crops like papaya, maize, and annatto. Cotton, pejibaye palm, arrow cane and gourds are others among the sixty or

Woman baking cassava cakes. The Yanomami have few material possessions. Their lightweight technology makes movement and travel easy. Hammocks are made of bunches of split vines. *(Victor Englebert/Survival International)*

so crops the Yanomami cultivate.

The crops grow quickly. Maize is ready for harvest after four months. The cocoyams and sweet potatoes are edible within half a year and the **cassava** tubers become ready for harvest after about nine months. Bananas and plantains become ready soon after. The pejibaye palms take longer to mature, only producing substantial harvests after three years or more. However, unlike most of the other crops, these palms continue producing, often for as long as a decade after the plots are abandoned. Figure 1 shows a typical Yanomami garden.

While the plants grow the Yanomami must diligently weed their plots to prevent their immediate reversion to forest. Nevertheless weeds flourish in all the gardens and after three years or so the effort of weeding becomes too laborious and the plots are abandoned and new ones cut. The reversion to forest then proceeds rapidly. It regains the appearance of primary forest after about fifty years, though it may take over a hundred for the full diversity of the flora and fauna to re-establish itself. The Yanomami rarely reuse forest that is less than 50 years old for gardens. Yet, in spite of this long fallow period, the population density is such that the Yanomami only need about 2% of their territory to maintain their gardening cycle.

The locations of their gardens determine the position of Yanomami communities. Local forest areas that are not too steep, not too rocky and not exposed to flooding etc are suitable for cultivation. As these areas in the immediate vicinity of the settlement become exhausted the effort of walking to the gardens and, particularly, carrying back the harvests, becomes gradually greater. If the average travel time between gardens and village increases by as little as $\frac{1}{2}$ an hour, the Yanomami may decide to move their huts to be nearer their producing gardens. Indeed,

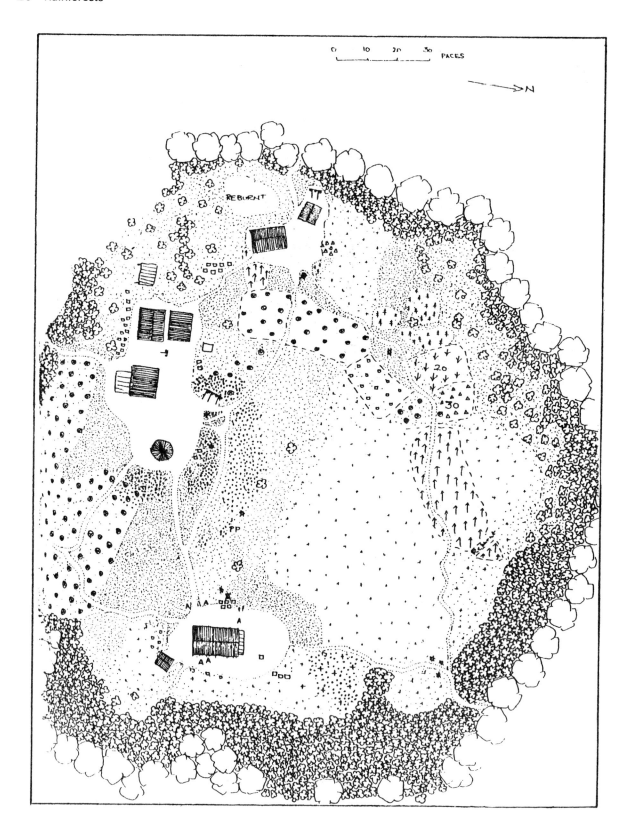

This sketch map of a typical Yanomami garden, shows the cropping pattern and layout of huts. Inter-cropping is limited.

Bananas and plantains are planted in wetter and lower parts of the garden. This mature garden has been heavily invaded by weeds. Quick growing softwoods, of the genus *Cecropia*, have already completely taken over the outer margins of the garden. Other parts of the garden have also already been given over to weeds. The garden was in fact completely abandoned about eighteen months after the drawing was made.

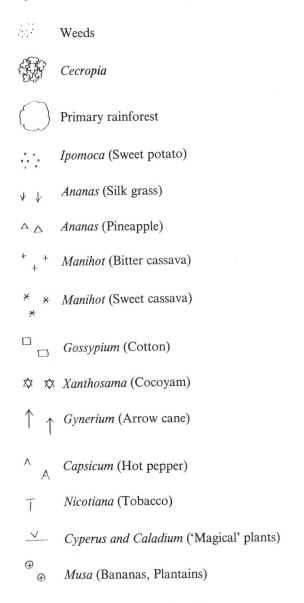

- Weeds
- *Cecropia*
- Primary rainforest
- *Ipomoca* (Sweet potato)
- *Ananas* (Silk grass)
- *Ananas* (Pineapple)
- *Manihot* (Bitter cassava)
- *Manihot* (Sweet cassava)
- *Gossypium* (Cotton)
- *Xanthosama* (Cocoyam)
- *Gynerium* (Arrow cane)
- *Capsicum* (Hot pepper)
- *Nicotiana* (Tobacco)
- *Cyperus and Caladium* ('Magical' plants)
- *Musa* (Bananas, Plantains)

Figure 1. Typical Yanomami garden

settlements may be moved small distances as many as three or four times a year, sometimes more, to take advantage of seasonal abundances in the various gardens.

Diet

The main staples grown in the gardens, bananas, plantains and **cassava**, have been selected by the Indians for their fast growing abilities and their capacity to do well on the very poor soils. These crops grow vigorously in the gardens and provide the Indians with a more than adequate supply of energy in return for their effort.

However, the soils, which are particularly low in nitrogen, offer little to more demanding crops like maize which produces poor yields and can rarely be grown more than once in the same plot.

For this reason, the staple crops provide the Yanomami with little in the way of protein, essential fats and vitamins and, though rich as sources of energy, do not provide a balanced diet *on their own*.

The Indians fully recognise the implications of their gardening for their diet. Their language distinguishes two kinds of hunger. Simple hunger, "ohi" satisfied with garden produce or starchy foods from the jungle and hunger for protein, "naiki". This second kind of hunger can only be satisfied with meat, fish, other small animal foods or certain rich and fatty fruits.

In order to achieve a satisfactory diet, to satisfy their "naiki", the Indians look elsewhere than their gardens for more nutritious foods – fruits, hunted game, fish and small animals gathered from the rainforest. By skilfully combining forest and garden foods they are able to achieve a very satisfactory diet with the minimum effort.

Gathering

The illusory luxuriance of the tropical forest disguises the fact that the nutrient cycles within it are tight and competitive. The forest is able to live on rather than off the soil. It achieves this by recycling nutrients direct from fallen leaf and bough to root and stem, through the abundant fungal hyphae which pervade the decaying layers of detritus and link directly to the great trees' rootlets.

Nevertheless, the prodigious rates of growth of the forest mean that it does sustain an abundant and sizeable animal population. Yet, these animals are, for the most part, small invertebrates and widely scattered mammals. To take advantage of them the Indians must adopt a similar strategy, keeping their own population density low and their villages widely dispersed. Gathering fruits and tubers from the forest and foraging for honey, edible insects, crabs, frogs and other minor game make important contributions to the Yanomami diet. Skilful at climbing trees and vastly knowledgeable in the ways of the forest, the Indians can subsist for long periods just by gathering forest products away from their gardens.

Fishing

The waters of the Guyana Highlands are as low in nutrients as the soils off which they flow. The small streams of the Sierra Parima do not abound with fish. Crabs and crayfish are important in the upland villages but fishing with the hooks and nylon lines, now available from the missions, make important additions to the diet and cushion the communities against protein shortage during poor runs of hunting.

The traditional and more productive form of fishing involves the collective use of fish poisons. Woody vines of the genus *Lonchocarpus* are cut from the forest and chopped to manageable proportions. These are then taken to the banks of small streams where they are first pulverised and shredded with wooden clubs and then dunked in the water, where they release their strong smelling poisonous white juices into the stream. The juices, which contain a mild nerve poison, rotenone, paralyse the fishes' gills allowing the asphyxiated fish to be readily caught in hand nets. When the conditions are right, substantial catches can be made in this way.

Hunting

The majority of Yanomami protein comes, however, from hunting. The Yanomami traditionally use bows and arrows for hunting. The bows are made from the strong flexible wood of palms, honed to shape using the lower jaw of wild pig. The bowstrings are made from the phloem tissue of young wild-growing *Cecropia* trees or of so-called "silk grass", a relative of the pineapple plant, cultivated in their gardens.

The Yanomami use long light-weight arrows made from the stems of arrow canes, which occur wild but which are also grown in their gardens. The arrows are notched, and fletched with the primary feathers of black curassows (so called bush turkeys). A variety of different points are used: barbed heads for shooting birds; needle sharp palm-wood points, notched and coated with arrow poison, for monkeys; long lanceolate heads made of carefully shaped bamboo for the large terrestrial mammals.

The Yanomami have three main types of hunt. At dawn and dusk individual men stalk with great stealth along the familiar paths in the immediate vicinity of their camps or villages. In the dim light, they rely on the crepuscular calls of birds, mainly quail, guans and tinamou, to give themselves away. Skilful hunters can accurately mimic the birds' songs to prompt them to call.

Most day hunting is also undertaken solitarily or in pairs, commonly the hunter accompanied by his wife. Such expeditions rarely take the hunter more than eight to ten kilometres from his village, the hunter trying where possible, to range through the forests about the community in an arc. Although sounds of game also play a key role during day hunting, visual clues are also of major importance. Hunters can recognise the spoor of most animals at a glance. By carefully examining the prints they can estimate how long ago the animal passed, whether it was moving slowly or fast and whether or not it was feeding.

Based on years of experience the hunters have learned, when it is worth following certain tracks and when it is not. Most important of all, they have learned how to move quickly and silently in their bare feet through the forests, slipping agilely through tangled lianas and past thorny stems and tree trunks, without stumbling or disturbing the game.

The most common animals taken during day hunting are birds, especially guans, tinamou and curassow. Macaws, parrots, toucans and doves are other common prey. Monkeys are also common

Man sorting out smoked meat. Although the Yanomami's gardens provide the bulk of their food, their crops are adapted to the poor forest soils and provide little protein. Meat from hunted game thus contributes an essential part of their diet. *(Victor Englebert/Survival International)*

game taken during the day hunt. The mobile bands of monkeys are frequently noisy and easily located, but also alert to intruders on the ground. Shooting monkeys high in the canopy of the forest as they bound from branch to branch requires great strength and an accurately aimed arrow. Poisoned arrow heads which break off in the wound are preferred, as the poisons act as muscle relaxants and prevent the monkeys dying still gripped to the trees.

The large ground mammals, while less frequently encountered and killed, nevertheless provide the bulk of the protein taken in hunting. **Agouti**, **capybara**, armadilloes, deer, tapir and collared **peccaries** are all regularly taken by the solitary hunter, though the latter three are now as often caught with the help of dogs, which have been introduced relatively recently to the Yanomami.

The most important prey species of all, however, the white-lipped peccary, is usually hunted collectively. Those out in the forests, whether hunting, foraging or collecting other forest products, who come across fresh prints of a herd of these gregarious peccaries will hurry back to the village with the news. Immediately or the next day, depending on the time and distances involved, the hunters will then sally forth as a group. After following the spoor, often for many kilometres, the Yanomami will attempt to surround the herd before moving in to the kill.

Frequently several animals are taken in a single hunt in this way. Such hunts take the hunters about as far as they are likely to go from a settlement in a single day, spreading the radius of hunted area around a community out as far as ten kilometres.

Trekking

Among the central and more traditional Yanomami, who have an even more casual approach to gardening than average, up to 40% of the time is spent away from the village on trek. Especially during the seasonal abundances of wild fruits and other forest produce, the communities divide up into smaller family groupings and move off into the forests, erecting adequate but quite temporary shelters when they pause in the afternoons to set up for the night.

While on trek the Yanomami may range much more widely from their villages, expanding the radius of their activities to between twenty and thirty kilometres about the community. Being away from the more intensively hunted and foraged areas around the communities, trekking is a time of abundant fruit and meat eating. But as the supplies of staple foods brought with them from the gardens dwindle and the labour of replacing them in the diet with tubers, palm hearts and starchy forest fruits becomes irksome, the Yanomami return to the more intense social life of the villages and the products of their gardens.

Affluence or poverty?

To gain all their foods, make their houses, gather their firewood, prepare their meals and weave their baskets and hammocks, spin their cotton and make their tools, the Yanomami, both men and women, work an approximately forty hour week. The rest of their time is their leisure, which they spend in raising their children, in discussing and organising their community life, in sleeping, resting and ritual.

Two dictums of the anthropologist Marshall Sahlins apply well to the Yanomami.

> 'There are two ways of being affluent, one is by acquiring more, the other is by requiring less................

> Since they must carry themselves the comforts that they possess, they only possess that which they can comfortably carry themselves'.

The Yanomami, through being almost wholly self-reliant and reducing their needs to a comfortable minimum, have learned how to sustain a kind of affluence from their difficult rainforest environment. Today, unfortunately, that independence is being undermined and their society with its subtle system of resource use is threatened with extinction.

2.5 Brief comparison with other Amazonian groups

The Yanomami are an upland people whose social organisation and systems of resource use have been cunningly adapted to suit the constraints of forest dwelling. Yet, their particular and distinctive society is but one of the many strategies that Amazonian Indians have used for maintaining themselves without destroying their forest environment. Other Indians have chosen different crops, have different social structures and use different techniques of hunting to achieve the same equilibrium.

Not all these differences can be explained in terms of differences in their tropical forest environment. However, the strong contrast between the Yanomami, other "forest Indians" and "river Indians" who inhabit the lowlands along the banks of the major rivers is quite clear.

The abundance of fish in rivers that have been mixed with the muddy nutrient-rich waters of the Andes has made the Indians in these areas much less dependent on hunting and gathering to make up their diets. Villages along the rivers are thus freed of the need to go off on trek for long periods. These settlements are at once more stable and larger and the gardens more substantial and carefully tended.

In some areas of the lower reaches of the Orinoco and particularly the Amazon, where the river floodwaters have created atypical soils that are quite rich in nutrients, the Indians developed intensive agricultural systems. These people relied on fast maturing crops of maize and beans, grown before the flood waters returned, to sustain large populations that developed complex technologies and social hierarchies. Turtle ponds and fish tanks made these people much less dependent on the immediate forests for their basic needs. Sadly these peoples were nearly all destroyed in the first centuries of contact, swept away by the ravages of diseases introduced by Portuguese and Spanish conquistadores.

The differences between "forest" and "river" Indians can be usefully compared in tabulated form.

Figure 2. Differences between "forest" and "river" Indians

Forest	River
Fishing is less important	Fishing important
Much hunting and foraging (long treks)	Less hunting and foraging (no trekking)
Settlements small and scattered	Settlements large and nucleated
Settlements mobile and unstable	Settlements more permanent and stable
Gardens small and not intensively farmed	Gardens larger and more carefully tended
Crops include manioc, cocoyams and bananas	Nutrient demanding crops such as beans and maize
Strongly egalitarian traditions	More emphasis on social hierarchy
No figures of authority	Well developed leadership
Minimal technology	Complex material culture

2.6 Amazonian land-use practices in transition

As the Indians are coming into contact with the outside world, their needs and problems are changing. Even when they escape the directly destructive effects of land invasion and new diseases, the Indians are nevertheless confronted with new pressures and presented with new opportunities.

In particular, the access of new technologies has created totally new possibilities in the use of their environments. Steel tools, which have now replaced the traditional stone axes throughout Amazonia, have made the labour of slash and burn agriculture about four times as efficient. Other industrial products quickly follow. Metal cooking pots and arrow points, shotguns, drills, needles, mirrors, scissors, beads, cloth, fish-hooks, nylon etc. all soon become integral parts of the Indians' economies, until suddenly the Indians find that they can no longer survive independently of the outside world.

To pay for these new necessities the Indians must produce a surplus, creating wholly new demands on what was previously a self-sufficient and self-limiting system. What begins as a simple transaction, machetes in exchange for manioc, axes for artefacts, sets in motion a whole process of social and economic change in the Indian societies. From being isolated, autonomous, independent, self-sufficient peoples with subsistence economies, the Indians became dependent, accessible and manipulable and linked to the market economy of western civilisation. The ineluctable process by which tribespeople become peasants is initiated. The change in orientation of the Indians' economies also has the effect of severely disrupting the relations with the natural environment.

2.7 The impact of imposed development

The disruption of Indian economies through their increasing adoption of the technologies and life-styles of outsiders is certainly not without problems. Nevertheless, many Indian communities have managed to re-establish a kind of equilibrium under such circumstances. The overall settlement patterns may have changed, with the small, mobile and dispersed communities giving way to larger, more sedentary villages nearer to the markets, but, if the overall population density does not increase too greatly, the general pressure on the environment does not pose too severe a risk. Eventually even the most settled villages will decide to move. The cycles of depletion and recovery, while longer, occur nevertheless.

26 Rainforests

Men sitting about chatting. Until recently the Yanomami's social universe was bounded on all sides by the forests. Although they had limited contact with other Indian groups their world was, to a large extent, a closed one. This isolation has now been shattered.

This is only possible however, so long as the land base of the Indians remains uninvaded and the *apparently* unused lands of Indian territory are still available for the villages to move to.

However, western concepts of land use, which evolved in the context of fixed plot agriculture in the temperate north, do not recognise the efficiency of the Indians' systems of resource use. 9 km² of land/person seems an awful lot of land by western standards. The lands are thus declared empty, the peoples vilified as obstacles to progress and invasion begins.

But as the case studies in this book show, the tragedy is that, to date, there are no "western standards" for the successful occupation of Amazonia. Introduced systems of land use have proved exploitative, destructive and unsustainable. Only an estimated 2% of the forest soils of Amazonia can sustain permanent agriculture. Experiments with cattle-ranching and agro-forestry have proved either ecologically ruinous or uneconomic. Yet the colonisation of Amazonia by the landless poor, displaced by agribusiness elsewhere in the continent, continues and is even accelerating.

The principal victims of this misguided venture have been the Indians themselves. Calculated at some nine million at the time of their "discovery", their numbers have been reduced in the following five centuries by some 90%. It is estimated that in Brazil alone one Indian tribe has vanished every year since 1900. Today the rate of invasion and extinction is, if anything, increasing as the roads cut deeper and deeper into the forests.

> *Activities linked to this chapter*
>
> Your teacher has details for the activities listed below.
>
> * General orientation to the Yanomami Indians (Activity 1).
> * Activities relating to Yanomami village micro-movements (Activity 2).
> * Activities investigating the extent of Yanomami hunting and trekking ranges (Activity 3).
> * Activities to demonstrate how industrial technologies can destroy the traditional Indian way of life and at the same time cause damage to the environment (Activity 4).

CHAPTER 3

The Extent of Deforestation – Amazonia and Beyond

3.1 Monitoring deforestation in Amazonia

Until relatively recently the inaccessibility of the Amazon rainforests completely defied anything other than speculation about the magnitude of deforestation. However, the use of LANDSAT satellite photography of the region has now dramatically improved the situation. Yet LANDSAT data are not without their problems. For a start, the high frequency of cloud cover in the Amazon makes LANDSAT satellite photography difficult at certain times of the year. Another problem is that LANDSAT cannot differentiate between primary rainforest and secondary forest that has grown up following rainforest clearance. It must also be noted that LANDSAT has only been in use since the 1970s and in many of the southern parts of the Amazon forest, extensive land clearing has been occurring for more than twenty years.

To date within the Amazon region, only Brazil has made the results of LANDSAT surveys easily available. Staff at Brazil's National Institute for Research in the Amazon (INPA) have analysed the most recent Brazilian LANDSAT data and the results have been published in a recent paper (Fearnside, 1986), the basis of this part of the book. Ideally, the following sections should be approached as a series of student activities.

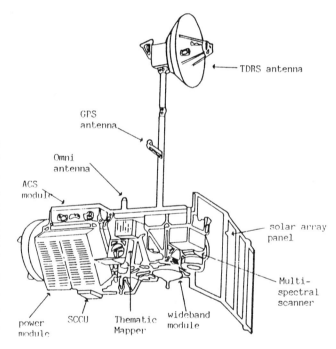

Multi-Spectral Scanner (MSS)

	Wavelength (μm)	Resolution
Band 1	0.50-0.60 (green)	80m
Band 2	0.60-0.70 (red)	80m
Band 3	0.70-0.80 (red-near IR)	80m
Band 4	0.80-1.10 (near IR)	80m

Thematic Mapper

	Wavelength (μm)	Resolution
Band 1	0.45 - 0.52	30m
Band 2	0.52 - 0.60	30m
Band 3	0.63 - 0.69	30m
Band 4	0.76 - 0.90	30m
Band 5	1.55 - 1.75	30m
Band 6	10.40 - 12.50	120m
Band 7	2.08 - 2.35	30m

Both sensors provide image data with 185 x 185km coverage, with 5.4% forward overlap and 7.3% side overlap at the equator increasing at higher latitudes.

Landstat 5

3.2 Deforestation pattern at the regional level in Amazonia

Deforestation in the Legal Amazon is shown in figure 3. Notice that LANDSAT data through to 1980 are available for only six covering 61.3% of the region's total area, of Brazilian Legal Amazon's nine states and territories.

The fact that LANDSAT images only became available in 1975 means that the value of the data at indicating long term deforestation trends is severely limited. A longer perspective can be gained by considering the extent of forest clearance in 1970 based on information from non-satellite sources. The cleared area was very small in 1970 as indicated by side-looking airborne radar (SLAR) images produced for the region in the early 1970s. In this INPA study the figure for 1970 is assumed to be zero. However, INPA points out that by 1970 secondary forest had formed in much of the long settled areas, e.g. around Belem. These would not be "seen" as "cleared" areas by either radar or LANDSAT, if it had been available.

3.3 Likely future patterns of deforestation in Amazonia

The rapid deforestation of the recent past leads INPA to believe that forest felling in Brazil's Amazonia in the coming years will also be rapid. The exponential increase will have to be modified eventually as the resources required to continue the trend to its logical conclusion become limiting. However, no smooth decrease or automatic end of present deforestation patterns can be predicted. Rather, infrastructure is now complete or nearing completion, so rapid spread of deforestation to more remote parts of the region has become easier. INPA believes that the foci of intense deforestation, now located mainly around the edge of Amazonia in Rondonia, Mato Grosso and southern Para can be expected to spread to other regions as access improves. The paving of the BR-364 highway, completed in September 1984 with financing from the World Bank's Polonoroeste loan, removed one of the major impediments to migrants.

The condition of the earlier version of the BR-364, in existence since 1967, grew steadily worse during the 1970s. In the rainy season it was common to see groups of over 400 trucks stuck in mud holes for periods of weeks or months.

A loan from the Inter-American Development Bank, approved in 1985, will finance paving the BR-364 from Porto Velho (the capital of Rondonia) to Rio Branco (the capital of Acre). Deforestation can be expected to increase dramatically when this impediment to population movement from Rondonia is removed. Similarly, government plans to improve the Porto Velho–Manaus highway should speed migration to Amazonas and Roraima.

Recent INPA data reveals that Roraima is currently in a phase of extensive deforestation due to migration from Rondonia. The best land in Rondonia has already been fully claimed and a growing stream of settlers who either arrive and do not find land, or who sell their lots to wealthier late arrivals, are heading for this new frontier. The government of the federal territory of Roraima is actively encouraging the flow and recently promoted it in a series of advertisements in Brazil's main weekly news magazines.

At the moment, Roraima appears to be inheriting Rondonia's role as the most rapidly growing frontier in Brazil's Amazonia. But this could easily be passed to other areas as yet relatively untouched. INPA predict that, within each of these areas, exponential clearing patterns can be expected during the early phases of immigration. However, it should be noted that after farmers have settled for several years in areas with well surveyed and documented farm boundaries, such as the older government settlement areas in Rondonia, there is evidence that clearing rates slow.

Philip Fearnside (1986) reaches the following conclusions about the deforestation situation in Amazonia.

1 At present, deforested areas in Brazil's Amazonia are small in relation to the huge size of the region. However, they increased rapidly during the period for which data are available (1975–80) and showed an exponential pattern over this period in several states.

2 Patterns change greatly as statistics are successively disaggregated from the level of the entire Amazon, to individual states or territories and to quadrats of one degree latitude and longitude or smaller.
3 Increase in deforestation in areas like Rondonia is dominated by immigration to colonisation areas. Even in the absence of high immigration rates, deforestation is proceeding rapidly in many parts of the region through the expansion of cattle ranching for land speculation purposes.
4 INPA investigations indicate that in older settled areas deforestation rates do slow down.
5 INPA expects an increase in the number of foci of deforestation as new "waves of exploitation" move from Rondonia to Roraima and later to other areas in the region.

3.4 Monitoring global deforestation – the UN Tropical Forest Assessment Project

Growing concern about the state of the world's tropical forests has led, in recent years, to a number of surveys each giving very different estimates about tropical forest *extent* and its present *rate of change*. To try to find out the true situation and provide an overall, objective assessment in 1978 the United Nations set up the "Tropical Forest Assessment Project" (Lanley, 1983).

Three major reports were published in 1982. These were inevitably large because this global survey was the first time much of the world's published and

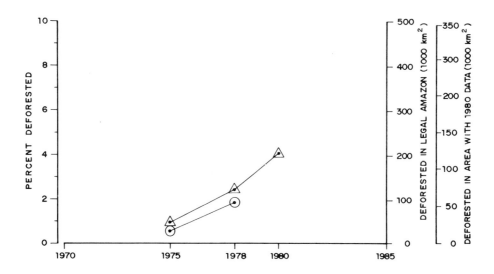

○ show complete data for all nine states and territories up to 1978. △ show states and territories with data complete up to 1980 (6 states only). Deforestation was considered to be zero in 1970 to facilitate visualization of the trend from 1970 to 1975.

Figure 3. Deforestation in Brazil's Legal Amazon (Fearnside, 1986)

People scurrying along highway. The shock of contact has led to a collapse of Yanomami society. Prostitution and begging as roadside nomads became a way of life for these Indians from the communities on the Ajarani river. Now 45,000 miners have invaded the Yanomami's territory. *(Adinair dos Santos)*

unpublished literature on tropical forests had been brought together. All this information was sorted out according to a single, uniform, classification system for tropical forest types. It has always been difficult to compare different scientists' deforestation estimates for tropical forests because of the confusion about which forest types were under discussion in any one survey. The UN's single classification system overcomes this problem and consequently provides, to date, probably the most reliable estimates about tropical forest extent and rate of change. It will serve as a baseline against which future changes may be measured. The United Nations plans continually to update the estimates with the aim of publishing tropical forest assessments every five years.

3.5 The UN tropical forest classification system

As figure 4 shows, the UN's classification system identifies two main forest types in the tropics, "closed" and "open" forests. It is fairly obvious that these main types need to be assessed separately:

1 "Closed forests": have an interlocking, continuous tree canopy, many layers and, usually, abundant undergrowth. They do not have a continuous dense grass layer which would allow grazing and the spread of fires. Closed forests are either broadleaved, coniferous, bamboo or mangrove and they occur in areas with high rainfall or locally abundant ground water. Note that "rainforests" fall within and make up most of the "closed broadleaved" category. Many other forest types fall into this "closed broadleaved" category but their geographical extent is much less than that of rainforest.
2 "Open forests": are mixed broadleaved forest and grassland formations usually without an extensive continuous canopy but with a continuous grass layer under a tree canopy that covers more than 10% of the ground e.g. "cerrado" in Brazil and the wooded grasslands of Africa. Such formations, in general, occur in regions that are drier than those supporting closed forests.

There is a third major forest type, though its extent is much less than the two main types.

32 Rainforests

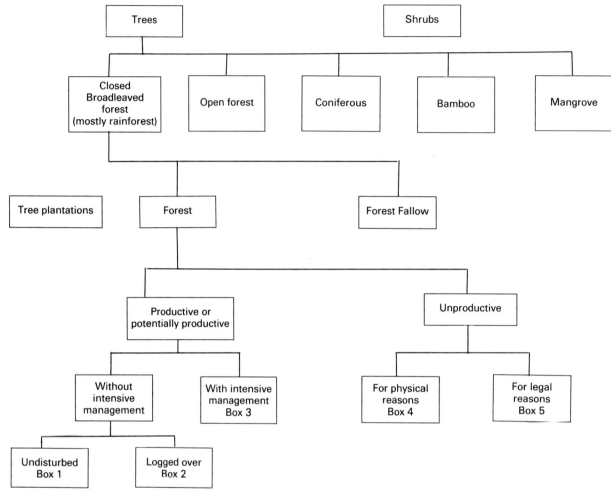

Note

* "Undisturbed forest" (Box 1) is forest which is, to date, totally undisturbed by human activity (except by tribal peoples), i.e. primary forest.
* "Logged over forest" (Box 2) is forest which has been commercially logged one or more times during the last 60 to 80 years.
* "Intensively managed forest" (Box 3). The UN has a very strict definition of "intensive management". It implies not only strict application of tree harvesting regulations but also silvicultural treatments and protection against fires and diseases. Unlike in "logged over forests", in these forests sustainable, "annual allowable cuts" are a reality.
* "Unproductive forest for physical reasons" (Box 4). These are forests which cannot produce wood for industry because of their characteristics, e.g. forests with stunted/crooked trees, or forests which cannot be logged because of terrain conditions, e.g. too rough or permanently flooded.
* "Unproductive forest for legal reasons" (Box 5). These are forests which are protected from commercial logging by law, e.g. designated as a national park or strict nature reserve.

Source: Adapted from 'UNEP/FAO Tropical Forest Assessment – a summary paper', J.P. Lanly, 1983

Figure 4. UN tropical forest classification system

The Extent of Deforestation – Amazonia and Beyond

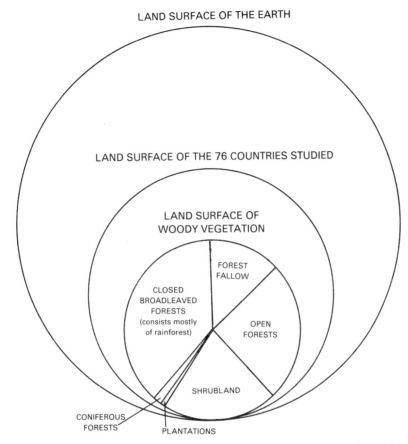

Note

* UN annual deforestation rates for *open forests* are: America, 0.59%; Africa, 0.48% and Asia, 0.61%.
* In *coniferous forests* the deforestation situation is extremely serious. Making up only 1% of the world's tropical forests, at the moment, they are the most threatened tropical forest type.
* Deforestation rates for *closed broadleaved forests* (rainforests) are considered in detail in Activity 7.

	THE AMERICAS		AFRICA		ASIA		THE WORLD
	Area in '000 ha	% of world	Area in '000 ha	% of world	Area in '000 ha	% of world	Area in '000 ha
Closed Forest	678 655	56.5	216 634	18.0	305 510	25.4	1 200 799
Open Forest	(216 977)	29.5	486 445	66.2	30 948	4.2	734 390
Fallow of Closed Forest	108 612	45.3	61 646	25.7	69 225	28.9	239 483
Fallow of Open Forests	(61 650)	36.2	104 335	61.4	3 990	2.3	169 975
Shrubland	145 881	23.4	442 740	71.0	35 503	5.7	624 124
							2 968 771

Source: 'UNEP/FAO Tropical Forest Assessment a summary paper', J.P. Lanly, 1983.

Figure 5. Global areas of tropical woody vegetation

3 "Shrubland": consists mainly of woody plants between 0.5 and 7 metres in height.

The remaining major classification units, shown in figure 4, relate to what is found in formerly forested areas after the forest has been cleared – in other words, *replacement vegetation.*

4 "Closed forest fallow": is derived from the clearing of broadleaved and coniferous forests for shifting cultivation. Fallow areas result after fields are abandoned. They usually consist of trees and shrubs at various stages of regrowth.

5 "Open forest fallow": consists of open forests in various regrowth stages after the original forest has been cleared for agricultural purposes.

6 "Plantations": are stands of trees established on land after the original forest has been cleared. Many tree plantations in the tropics consist of a single tree species (monoculture), e.g. eucalyptus species are often planted. In most respects, plantations do not replace natural, undisturbed forest. However, by providing an additional source of wood for industrial and domestic purposes, they do reduce pressures on undisturbed forests.

3.6 UN deforestation estimates in tropical forests

The data and diagram in figure 5, show the extent of the world's tropical forests. Some key statistics concerning deforestation rates in "open forests" are provided in figure 5. It is important to realise that *"rainforests" fall within and make up most of the "closed broadleaved forests" category.* Regrettably, it is beyond the scope of this text to look at the open forests situation in more detail.

Instead, *the student activities accompanying this section of the book focus only on rainforests.*

Activities linked to this chapter

Your teacher has details for the activities listed below.

* Investigations into the spatial variation of deforestation in Brazil's Amazonia (Activity 5).
* Activities relating to land uses in rainforests in different parts of the world (Activity 6).
* Calculations of deforestation rates in rainforests in different regions (Activity 7).

CHAPTER 4

Impact of New Rural Development

4 ... recent ... – an ... (based on ... Branford and Glock, 1985)

In the last century the patterns of human colonization and forest destruction in Brazil's Amazonia have been produced by at least five inter-related thrusts of economic/land use activity.

These are:

- recent pioneer smallholder colonization
- cattle ranching
- infra-structure development
- energy development, hydro electric power
- mineral extraction.

It can be useful to envisage these different activities in Amazonia as "migratory flows", with impacts on the region's natural resources varying over the years, depending on the number of people involved, intensity of land use etc. Many factors affect these different migratory flows as demonstrated in the case of pioneer smallholders in section 4.3 of this chapter.

As Johnson (1985) argues these 5 key economic activities all have national security, socio-political and ecological, as well as economic dimensions. All inter-lock.

It is particularly important to appreciate the national security dimension. Since the mid-1960s Brazil's military leaders have been closely involved with the official Amazonian development projects.

One fundamental geo-political consideration has motivated leaders to seek development of the Amazon Basin since early in the country's history: the fear that, if Brazil does not occupy the Amazon, somebody else will. "Occupar Para Nao Entregar" has been the motto of military leaders and politicians who endorse this justification. According to Mario Andreazza, former Minister of the Interior, this was the main consideration behind plans for the Transamazon Highway (see figure 6). Other observers would strongly dispute Andreazza's viewpoint. The strategic highway concept, it is claimed, was only used as a justification of the highway after the event.

Significantly, this is not only the concern of politicians and the military. The "Commission in Defence of the Amazon", a diverse coalition of former military men, professionals, some politicians and environmentalists, combines nationalistic concern over control of Brazil's Amazonia with opposition to foreign exploitation and a desire to protect the ecological integrity of the rainforest. This coalition received its major impetus from a 1967 Hudson Institute proposal to dam the main stem of the Amazon river system in order to create a gigantic source of hydro-electric power and form a great lake in the centre of South America (visions of the Great Lakes!) to promote transport and trade. The plan was attacked as a plot to enable the USA to establish a submarine base, and, surprising though it may seem to foreigners, it is still cited by many Brazilians as a prime example of the danger of letting multi-national corporations into the Amazon.

Road through forest. *(WWF)*

Brazil's sensitivity over national identity and security in relation to its immense, and still virtually unexplored, rainforest territory has not been confined to the proposals of think tanks such as the Hudson Institute. Brazil's President Figueiredo was apparently sceptical of Peruvian President Belaunde's river-linking scheme of the late 1970s for similar reasons. Belaunde had proposed a mega-project of canals and dams which would link the Orinoco river in Venezuela with the Brazilian and Peruvian Amazon in the centre of the Paraguay and eventually the Plata in the south. His strategic ideas, however, made many Brazilians uncomfortable.

Thus, at the national security level two powerful forces always tend to act against each other. On the one hand there is the wish to occupy Amazonia's interiors; a potent motive for regional occupation and resource exploitation. On the other hand there is the mutual suspicion of the countries of the region. This, of course, frustrates the type of joint project envisioned by Belaunde.

4.2 Recent pioneer smallholder colonization, cattle ranching and infrastructure development

From the beginning of the 19th century an increasing number of poor, small-scale cattle-rearers had begun to move into Amazonia from the east. This front moved slowly across Maranhao (see figure 7). In the first decade of this century this migratory flow was boosted by poor farmers from northeast Brazil, and very soon this front reached the Tocantins River. The wave of migration gained impetus after the collapse of the rubber boom and some families gradually advanced into Para in the 1930s and 1940s, often forced on by the exhaustion of their farm-plot soils.

With the worsening agrarian crisis in the northeast in the 1950s, hundreds of thousands of families migrated south to the big industrialized cities e.g. Sao Paulo. It was the beginning of the famous "urban

Burning forest. *(J. Etchart/Reflex)*

explosion". But thousands of others took a less publicised route, travelling west into Maranhao in the wake of earlier Amazonian migrations. During this period, migrants from the northeast also moved into Goias for mineral prospecting.

As a result of this large influx of people, the available land soon began to run out in Maranhao. Events in Goias evolved in a somewhat different fashion. Farmers from the south of Brazil had been attracted to the state since the beginning of the century. However, from the 1950s, with increasing land pressure in Sao Paulo and Minas Gerais, the flow to Goias increased. This led to a redistribution of the state's rural population with more modern, better-off farmers from the south buying out long-established Goias farmers.

From the 1950s many poor pioneer smallholders began to push further west, crossing the Araguaia River into north Mato Grosso. There is a marked difference today between the older communities to the east of the Araguaia River in the states of Goias and Maranhao, which were established during the slow, peaceful migrations of the first half of this century, and the more recent settlements to the west of the river in the states of Para and Mato Grosso. The latter were largely formed during the faster and more violent waves of occupation in the 1960s and 1970s.

The occupation of Goias was greatly accelerated by the construction in 1960 of the 2000 km Belem-–Brasilia highway, the BR-153 (see figure 6). The building of the road marked the beginning of a new phase because, for the first time, the road network became more important than the rivers and cattle tracks as the main route for poor families migrating into Amazonia. The building of the road was also the first government attempt to open up the region. The population inflow around the Belem–Brasilia road was very large and it has been estimated that, between 1960–70, 170 000 people moved into the region,

Recent settlers' town. *(M. Santilli/Panos)*

occupying the land on either side of the road. Goias state population statistics are shown in figure 8. The construction of the Belem–Brasilia highway also marked the beginning of the phase of very violent confrontations in the fight for control over the land in the region. For the first time, the road gave Amazonian farm produce access to outside markets and this meant the value of the land rapidly increased.

1964 marked the end of two decades of civilian rule in Brazil and Branford and Glock (1985) argue that the military rulers that took over began to see the occupation of Amazonia as an important objective for two main reasons:

- as mentioned earlier, they wanted to achieve the military goal of territorial occupation.
- second, the military leaders believed that the migration of poor families into Amazonia threatened to destroy and degrade the region's abundant natural resources. They believed most of the land should be occupied by wealthy and "sophisticated economic groups capable of rational exploitation of the region".

Thus the military government devised a two-pronged strategy to promote the type of occupation it thought desirable. First, it worked out ways to attract wealthy groups into the region. The system of government subsidies, agreeing to cover the initial losses of big cattle companies, became a crucial incentive.

In the second place, the military government undertook an ambitious road building programme in Amazonia in the late 1960s and early 1970s, see figure 6.

Some authorities believe these roads were primarily built to allow the big cattle companies to move in. Certainly, as a result of the road network, the population of the Amazon more than doubled from 1960 to 1980 with profound effects on the region.

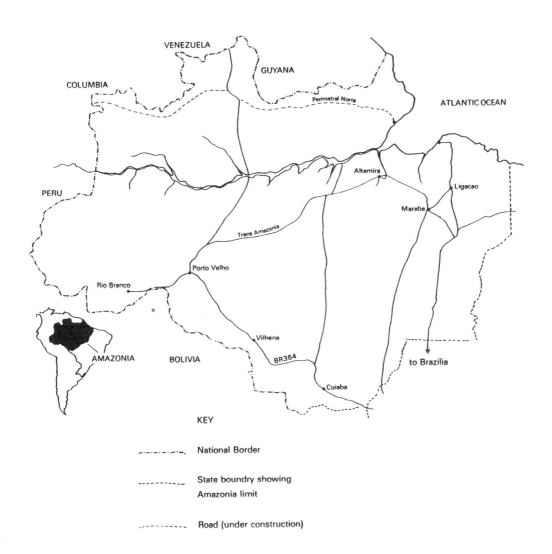

Figure 6. Main road network in Amazonia

Figure 8 shows the variation in timing of the migratory flows in different states. This huge investment in infra-structure has also certainly resulted in escalating land conflict in Amazonia – evidence of a rising number of violent conflicts and land evictions during the 1970s has been documented (Branford and Glock, 1985). Experience in the region shows that, when a road is built many pioneer smallholders arrive and they are accompanied by market-orientated entrepreneurs, the cattle ranchers, land-speculators and mining companies.

Whereas in the 1950s and 1960s nearly all the families moving into the Amazon came from the northeast, in the 1970s more and more of the migrants came from southern Brazil. During this later period the state of Rondonia became a major destination for many of the pioneer smallholders (see figure 7).

Note
Arrows do not indicate size of migrations.

Figure 7. Main migratory flows during the 19th and 20th centuries (Branford and Glock, 1985). Reproduced with kind permission of Zed Books.

		Total population		
	1950	1960	1970	1980
Goiás	1 214 921	1 913 289	2 938 677	3 864 881
Mato Grosso	522 044	889 539	1 597 090	1 141 236
Mato Grosso do Sul[a]				1 368 803
Pará	1 123 273	1 529 293	2 167 018	3 411 235
Amazonas	514 099	708 459	955 235	1 430 314
Rondonia	36 935	69 792	111 064	492 744
Acre	114 755	158 184	215 299	301 628
Roraima	18 116	28 304	40 885	79 078
Amapá	37 477	67 750	114 359	175 634
Total	*3 581 620*	*5 364 610*	*8 139 627*	*12 265 553*

[a] Mato Grosso do Sul was created as a separate state in 1979; until then it formed part of Mato Grosso.

Source: *Anuário Estatistico do Brasil* (FIBGE, 1980).

Figure 8. Population of the Amazon region, 1950–1980

4.3 Factors promoting migration of pioneer smallholders into Amazonia

Complex and inter-related factors are at work. Migration into the state of Rondonia beginning in about 1970 is a good example of this. Many of the factors promoting migration to Rondonia apply to earlier migrations into other parts of Amazonia. Factors promoting migration include:

1. The completion of the BR-364 road in 1965. The road deteriorated and remained closed for weeks during each rainy season, but in 1969 the road was improved.
2. Many of the coffee and food crop lands in the southern and eastern states of Brazil are being turned over to mechanized export crops like soybean and vast plantations of sugar cane for alcohol production. Result: reduced rural employment.
3. Since 1970 heavy frosts have killed large areas of coffee in Sao Paulo and Parana. The result is reduced employment.
4. Droughts in the northeast of Brazil.
5. The plans by Rondonia's state government and INCRA to allot large, usually 100 ha, agricultural holdings to families migrating into the state. The rapid spread of compelling, and largely erroneous, rumours of fertile soils in Rondonia.
6. The abandonment, in 1974, of official attempts to settle large numbers of people along the Transamazon Highway in favour of agro-industrial cattle ranching. This was only 4 years after construction began.
7. Landlessness: the large influx of migrants into Rondonia and other parts of Amazonia is to a large part, due to the highly concentrated system of agricultural land tenure in Brazil, one of the most skewed in the world. According to Brazil's 1975 agriculture and farm husbandry survey, only 0.8% of all Brazilian farms covered 1000 hectares or more, but they accounted for 43% of the occupied area. At the other end of the scale, 52% of the farms were 100 hectares or less in size and they accounted for just 3% of the land.
8. High population growth rates in many areas. This has often led to reductions in farm sizes as families subdivide their plots among their children. Eventually farm size becomes unviable, resulting in migration.
9. Huge overseas investment by the World Bank in Rondonia since 1981. Since 1980, migration rates into Rondonia have doubled and there is little doubt that World Bank involvement in the Polonoroeste Programme has been a key factor promoting this.

One of the major reasons why the Polonoroeste Programme, in the state of Rondonia, has been selected as a case study in land use in the Amazon is that it is a clear demonstration of how countries in the North often play a direct role in the destruction of the environment in developing countries. In the case of Polonoroeste, *the direct investment of World Bank capital* is a key factor "pulling" migrants into Rondonia with resultant large scale rainforest destruction. In chapter five World Bank links with rainforest destruction are examined in considerable detail.

4.4 Pioneer smallholder annual crops

The thousands of pioneer farmers, both *squatters* and *government sponsored colonists* who have entered the Amazon in recent years from other regions employ agriculture with many essential differences from the traditional long-cycle shifting cultivation of Amerindian and caboclo populations. The new arrivals fell and burn the forest, much as do traditional shifting cultivators in the first step of a swidden cycle, but thereafter the differences in the 2 systems becomes more apparent.

A few of the pioneers are from caboclo backgrounds in other parts of Amazonia; these people carefully select the land to be cleared based on indicator tree species present, and plant a diversified array of crop plants. They are also more skilful in timing the felling and burning operations to obtain the best burns, as well as making the many agricultural decisions from deciding how much to plant of crops like rice, requiring intensive periods of seasonal

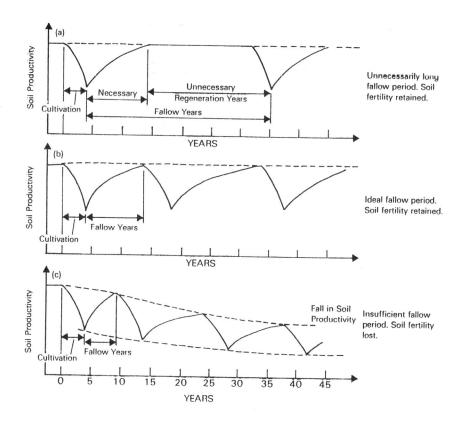

Figure 9. The relationship between deforestation and crop failure (after Ruthenberg)

labour; as compared with more traditional staples like manioc, which spread the labour requirements over much of the year.

However, it is widely accepted that most new arrivals from other ecological regions, like the northeast and south, find adaptation to the new environment difficult. Many gradually adopt some of the solutions long practised by the area's residents. The speed and path of the adaptation process varies greatly however, depending in part on the colonists' background before arrival. Pioneer farmers do not plant the wide variety of crops employed by traditional shifting cultivators. They primarily grow rice, manioc, corn and various beans.

The most striking difference between pioneer agriculture and traditional shifting cultivation is lack of the cultural tradition which leads swidden farmers to leave their fields in second growth ("bush fallow") for long periods before returning for a subsequent crop. Pioneers clear young second growth only one or two years of age with high frequency, not a practice that could be expected to continue for long without serious decline in soil productivity.

The main causes of pioneer crop yield declines are: decline in soil fertility; relentless competition with weeds; rapid pest build up and an increasingly harsh microclimate. The relationship between deforestation and crop failure is shown diagramatically in figure 9.

The Impact of New Agricultural Development   43

Settlers planting fields. *(M. Santilli/Panos)*

Philip Fearnside (of Brazil's INPA) points out that it must be realised that, from the outset, most colonists have no intention of using a sustainable cycle of shifting cultivation as the basis for their agriculture. Rather, annual crops planted in the early years of settlement are seen as a temporary solution to their immediate needs for cash; while the settler waits for a change to other sources of income such as cattle pasture, perennial crops, or selling the land at a good price to someone else who will develop one of these longer-term uses. By far the greatest share of the land area, both in areas inhabited by small colonists and in large land holdings, is rapidly being converted to cattle pastures.

Pasture is the most common land use in terms of area, even in settlement zones that have been the focus of the most intense promotion of perennial crops. In the Ouro Preto colonisation project in Rondonia, for example, 100 lots surveyed in 1980 had 49% of the cleared area under pasture. Much of the recently-cleared land eventually tends to be planted to pasture after an initial period of time in other land uses. In the much larger areas of the Amazon Region, where large cattle ranches rather than small colonists predominate, the fraction of cleared land going into cattle pasture is closer to 100%.

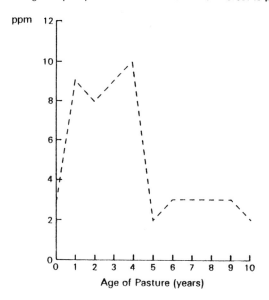

Figure 10. Changes in phosphorus after conversion from forest to pasture

4.5 Cattle ranching

This is the most important agricultural activity in Amazonian rainforest. According to Philip Fearnside, it is growing at a high rate and will dominate ever-increasing parts of the Amazonian landscape. Beef productivity is low, but, far more importantly, it is unlikely to prove sustainable. Studies show that the dry weight of pasture grasses produced per hectare per year is small largely due to poor soil. Available phosphorus has been found to limit grass yields in several locations on typical soils. Pastures are also quickly invaded by inedible weeds, better adapted than the grasses to the poor soil. Of an estimated 2 500 000 hectares of planted pastures in the Brazilian Amazon in 1978, 20% were considered "degraded" or invaded by second growth.

Phosphorus is the most crucial element for pasture production in the Amazon and 100 ppm is usually considered the minimum value for sustained production of pastures. After forest clearance, soil P values increase dramatically due to burning, but, by the fifth year, studies show they drop to about 2 ppm and steadily decline thereafter. A graph of P changes in pastureland is shown in figure 10.

Phosphorus decline has been identified as the major reason for pasture instability in the Amazon. The high demand of the grass *Panicum* for this element, coupled with losses due to erosion and animal export in the carcases, and the competition the grass experiences from low P adapted weeds leads to drastic drops in pasture productivity, which often results in pasture abandonment. Studies show *Panicum*'s excellent response to phosphorus fertilization, but the high transport and application costs, coupled with erratic availability of P fertilizers in much of Amazonia makes widespread pasture fertilization uneconomic at this time. Also note that soil compaction, leading to reduced infiltration and increased erosion, low seed viability of *Panicum* in Amazonia because it is an African savanna species and other problems affecting sustainability have been recorded on Amazonian cattle pasture.

It is important to appreciate that most cattle ranchers in Amazonia care little about whether they produce beef efficiently and sustainably! They are frequently just land speculators and use ranches to "put their mark" on the land and wait for its value to increase. As more and more roads are constructed and Amazonia's minerals, forests and other resources become more accessible, speculators expect the future prices of the land to rise. Aided by government ranching subsidies, which help minimise initial financial losses, these people see ranching as the best way *physically* to occupy and hold a large land area with the minimum labour and financial expenditure. Ranches in Amazonia employ very few people per unit area. Gaining physical occupation is crucial; Amazonia's land tenure system is based almost entirely on physical possession of land. Formal documentation of land titles normally occurs after the "owner", or his representatives, have occupied the area. Because the land is so valuable, land speculators frequently use violence and fraud to eliminate less powerful competitors for claims, especially small farmers and Amerindians.

It is not only the threat of violence which induces small farmers to sell up their land to ranching interests. Even though pioneer settlers might not view themselves as land speculators they are tempted to sell their plots to receive a relatively large lump sum; a reward which is usually much more than they have ever made through their farming efforts. Clearly, the lure of speculative profits hinders sustainable agricultural development by raising land prices to a point where most productive small farmers are excluded.

4.6 Other agriculture on terra firme

Small areas are presently devoted to: perennial crops like cacao, coffee and rubber and the mechanized cultivation of annual crops like rice and beans; horticulture.

4.7 Farming on varzea

Amazonia's **varzea** (floodplains) has been much used by Amerindians in the past. They have been used to a lesser extent by caboclos in the last 100 years. While varzea soils are more fertile than almost all **terra firme** soils, the most important difference is not that varzea yields are higher when a crop is harvested. Far more

For smallholders near Amazon rivers fish is a major source of protein. *(Oxfam)*

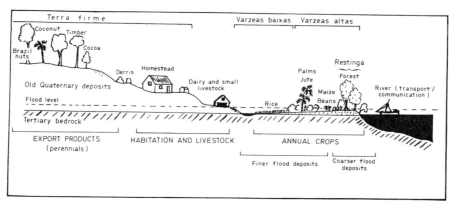

Figure 11. A varzea utilisation scheme (Barrow, 1985)

A varzea utilisation scheme, based on proposals made by F. de Camargo in the 1950s. During floods the velocity of the silt-laden water decreases away from the river channel, heavy particles settle first, and furthest from the river where flow is slow fine sediment is deposited – the result is a gently graded levee (vertically exaggerated in the drawing). Coarse soils form the higher "varzea-alta", and less well drained, softer soils form the lower "varzea-baixa".

significant is the possibility varzea offers for a sustainable yield without the lengthy fallows, or heavy fertilization, required to maintain annual crop productivity for more than 2 or 3 years on terra firme. Annual flooding in the varzea deposits a fresh layer of fertile silt and leaves the land virtually free of weeds and pests, at least once a year, at the moment when the river flood water recedes. At the moment mechanized cultivation of irrigated rice is being pursued in one location on the varzea as part of the Jari Project. But note the irrigated rice at Jari fails to take advantage of the varzea's major advantage which is annual renewal of soil fertility by flooding and siltation. Zebu cattle are grazed on natural varzea grassland. Water buffalo are of increasing importance being well adapted to the wet conditions of the floodplains. A varzea utilization scheme is shown in figure 11.

4.8 Human carrying capacity of land, land tenure distribution and human consumption patterns

Any discussion of sustainable agricultural land uses needs to consider "human carrying capacity" which can be defined (Fearnside, 1985) as "the density of people that can be supported indefinitely in an area at an adequate standard living without environmental degradation, given appropriate assumptions concerning productive technology, human consumption habits and criteria for defining an adequate standard of living and acceptable environmental degradation".

The question of land tenure distribution also lies at the root of any discussion of agricultural land use and production in Amazonia.

Activities linked to this chapter

Your teacher has details for the activities listed below.

* Identifying factors promoting migration into Amazonia (Activity 8)
* An activity looking at the global interdependence of Amazonia (Activity 9)

CHAPTER 5

The Impact of Regional Plans: Agriculture and Industry

5.1 Regional plans for Amazonia: an introduction

It is important to look at the Brazilian government's plans for Amazonia in the context of the country's massive international debt. Presently standing at US$ 120 billion, it places immense pressure on the government to exploit the country's natural resources, including the minerals, hydro-electric power and other resources in Amazonia. The newspaper extract from 'The Guardian' highlights the drastic measures recently introduced in an attempt to overcome the crisis facing Brazil's economy.

5.1.1 Energy development – Hydro-electric power

The hydro-electric power (HEP) potential of Amazonia is enormous, although the total is unknown. It is estimated to exceed 85 000 MW. In late 1984, the giant Tucurui dam on the Tocantins River was completed, starting the creation of a 2 200 square kilometre reservoir with potential for 8 000 MW. Tucurui is simply an early stage of a $60 billion plan to exploit the mineral resources of the Carajas Mountains (see figure 12). The dam will provide the power to run copper and aluminium smelters and to feed into the national grid. The Brazilian government sees this whole Carajas region as the Amazon model of the future which will integrate farms, factories, towns, highways and railroads in a pattern that many planners hope will multiply right across the rainforest.

Tucurui is the first step in this grand strategy. Among other major HEP schemes on which work has already begun are Balbina, a dam near Manaus, and another, Samuel, near Porto Velho in Rondonia. Plans are in preparation for a dam larger than Tucurui on the Xingu River, the Amazon tributary to the west of the Tocantins.

HEP schemes can be environmentally sound, representing a way to harness energy on a sustainable/renewable basis. However, large-scale dam projects in recent decades demonstrate that there can be high environmental and social costs, such as the forced resettlement of people from the flooded zone, the loss of habitat, the risk of water-borne diseases, and so on. The Aswan Dam in Egypt, for example, does provide many benefits but it has also created numerous social and environmental problems such as lack of silt to renew soil fertility, the spread of water-borne diseases, and a decrease in Mediterranean fish catches.

In Amazonia the future challenge will be to minimise the costs of HEP schemes and ensure that the power produced provides real benefits to the majority of the population. There is some doubt as to whether the challenge can be successfully met. On the other hand, many environmentalists and development planners point to the "cleanness" of HEP in relation to other energy alternatives, particularly nuclear.

Brazil leader pledges sweat and sacrifice in war on inflation

Jan Rocha reports from Sao Paulo on President Sarney's final gamble to recover credibility ahead of the election

PRESIDENT José Sarney has made a last gamble to recover some credibility and carry Brazil through to the November elections without succumbing to the chaos and upheaval of hyperinflation, declaring an all-out war on inflation, and promising sweat and sacrifice.

The success or failure of the President's initiative will directly effect the outcome of the election.

In a nationwide television and radio broadcast, the President made an apocalyptical forecast about the future unless his plan of "national salvation" succeeds in reducing inflation.

He was speaking to an estimated audience of 80 million listening and watching him from the Amazonian river banks to the slums of Sao Paulo.

Calling them a huge salad of drastic new measures — "the toughest ever introduced" — they include devaluation, deindexation, a new currency, a prices and wages freeze, a credit squeeze, government austerity and tighter taxes.

He also threw in the government reshuffle, closing down five ministries and 42 state companies, and said that from now on government will not spend a centavo more than it earns in revenue.

Sixty thousand civil servants are now threatened with redundancy.

With three zeros lopped off the cruzado, which now becomes a New Cruzado, overnight billions have become millions, millions have become thousands and one thousand old cruzados are now worth just one New Cruzado.

This is the third ambitious anti-inflation plan in four years announced by President Sarney, whose government nonetheless has accumulated a record total of 17,000 per cent inflation.

The new plan, baptised the Summer Plan because of the season here, was not greeted with any of the euphoria of the first plan in 1986. There was good cause for scepticism. News of the changes had been leaked days before, shops and supermarkets had already embarked on a frenzy of price mark-ups ready for the freeze. The government itself readjusted fuel, petrol, postal, electricity, telephones, milk and bread prices on the eve of its own announcement.

In the opinion of some economists the new plan will bring recession. Others see it as a preliminary to an important change in Brazil's position on the foreign debt, now standing at US$120 billion.

After the defiance of Brazil's 1987 moritorium, the Finance Minister, Mr Mailson Da Nobrega, had restored Brazil to the good books of the creditors by restarting interest payments and negotiating a new agreement. But repayments are crippling Brazil's capacity for internal investment and has not produced a return in the form of new loans.

The minister hinted while talking about the plan, that "Brazil will not hesitate to suspend interest payments if necessary for maintaining foreign currency reserves."

All exchange operations have now been centralised in the capital's central bank, an essential preliminary to the suspension of repayments. For some, this means a new moratorium is around the corner, while for others it is Brazil's way of putting pressure on the new American administration for a political solution to the debt problem.

The new measures still have to be approved by Brazil's Congress which is being called back from its recess. Politicians, with one eye on the November elections, are likely to adopt a wait and see attitude to gauge public response before they decide what to do.

Originally the President planned to introduce the measures together with a sweeping ministerial austerity programme, reducing the Cabinet from 27 to nine ministers, to be led by respected newcomers to the government.

But no eminent newcomer wanted to join such a discredited government and the President was reduced to simply downgrading five ministers.

Among those to go is that of land reform, marking the end of any serious attempt to change Brazil's hopelessly concentrated land structure.

'The Guardian', January 19, 1989

5.1.2 Mineral extraction

The enormous *known* mineral wealth of the Amazon Basin, apart from oil in the western Amazon, is largely concentrated in the Serra dos Carajas, a group of ridges and plateaus 160 kilometres southwest of Tucurui. The Carajas mountains contain one of the two or three largest iron ore deposits in the world. With 66 per cent iron content, the deposits are of a very high grade.

Since 1980 the Brazilian government has committed itself to an all-out effort to develop Carajas. Since then the discovered mineral riches of the area have continued to grow. According to estimates from the Brazilian national mining firm, other valuable minerals that have been found include: manganese deposits estimated at 60 million tonnes; copper, 1 billion tonnes; bauxite, 40 million tonnes; nickel, 47 million tonnes. Brazilian planners make no secret of their

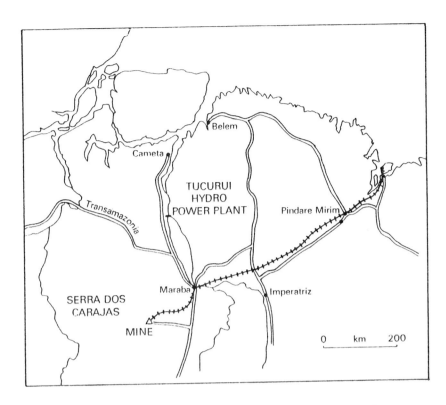

Figure 12. Map showing Tucurui hydro-electric power project and Carajas minerals programme

belief that the Carajas region is 'the treasure chest that is going to pay for putting Brazil into the 21st century'. The first commercial loads of ore began to move in 1985. Already a 900 kilometre railway through the rainforest is complete, connecting Carajas with a new deep water port at Sao Luis on the Atlantic coast (see figure 12).

US$ 5 billion has already been invested in iron ore extraction and movement to the Atlantic. There is foreign financing of which 33% comes from the EEC, the World Bank, Japanese and West German sources. With increasing political and economic uncertainty in South Africa, it seems that the strategic importance of Carajas to the world economy can only increase.

Interestingly, the environmental impact of Carajas and the Tucurui HEP project has been compared to that of *Polonoroeste*, a huge Amazonian agricultural settlement programme considered in detail in this chapter. Goodland (1985) points out that in the case of Polonoroeste, 100 000 square kilometres of primary rainforest may eventually be destroyed. This compares with 2 000 square kilometres at Tucurui and less than 100 square kilometres at Carajas. From this point of view, the exploitation of a concentrated resource, very small in area, as at Carajas, is preferable to large-scale forest conversion as at Polonoroeste.

50 Rainforests

Unloading of iron ore at railway terminus at Carajas for transportation to Sao Luis. *(T. Hall/Oxfam)*

5.2 Polonoroeste and the World Bank

Over the last 20 years the World Bank and other Multi-lateral Development Banks (MDBs) have financed hundreds of agricultural projects in the world's rainforests. In Latin America the focus has been on cattle ranching projects. Until recently the World Bank had no investments in Amazonia. However, in mid-1982, the Bank appraised the Amazonas Agricultural Development Project which focused on agriculture in the seasonally flooded varzea zone mainly south and east of Manaus. Now the Bank's major Amazonian investments are in the Polonoroeste Programme, the agreement was signed in December 1981, and the Carajas iron ore project, signed in August 1982.

The Northwest Region Integrated Development Programme (Polonoroeste) is a huge land utilization scheme for the Brazilian states of Rondonia and parts of Mato Grosso and is, to a large extent, funded by the World Bank. This chapter provides a compilation and analysis of materials from various sources which we hope will encourage you to examine critically whether Polonoroeste has caused environmental and social impacts in Brazil's rainforests.

After briefly describing Polonoroeste and its objectives, this chapter examines the many criticisms about the programme. The World Bank responses to these criticisms are also considered. The environmental impact of World Bank projects in the Third World have recently been closely scrutinized by the US Congress. These Congress investigations and the Bank's responses are considered in some detail.

Undoubtedly the issue of Multi-lateral Development Banks' environmental and social impacts is now very much on the international environment-development agenda.

However, it is important to appreciate that the World Bank and other MDBs are not the only institutions which invest large amounts of capital in developing countries. Some people argue that many of these other institutions also have investments which lead to major environmental and social problems.

5.2.1 Recent history and description of Polonoroeste

The Northwest Region of Brazil is officially defined as the area of influence of the 1 500 kilometre Cuiaba – Porto Velho (BR 364) road, thus encompassing all of the state of Rondonia and the central and western parts of Mato Grosso (see figure 13). This 410 000 square kilometre area, about three-quarters the size of France, is sparsely populated with 1 200 000 people, less than three inhabitants per square kilometre.

In the early 1900s, Marshal Candido Rondon joined Cuiaba with Porto Velho during construction of the telegraph, and an earth road was built in the 1960s. The road deteriorated and remained closed for weeks during each long rainy season and, even when passable, entailed very high transport costs. Distance, difficulty of access, and an environment less attractive to settlers than more accessible areas in Brazil, kept the region relatively undisturbed until 1970.

Since then the situation has changed dramatically and numerous factors have resulted in large migrations to the Northwest Region, see chapter 4. Goodland (1985) reports that by the late 1970s, the region 'was one of the most dynamic in Brazil. With 5 000 people arriving per month with few financial resources, little employment, and nowhere to go, the region became volatile and unhealthy, if not dangerous'.

In 1979 the Brazilian government invited the World Bank to survey the region, to assess land use/development potential, and to identify issues concerning possible financing. Following the 1979–80 survey, the Bank decided to help finance part of the Polonoroeste programme. The Polonoroeste programme is a group of projects budgeted at US $1.55 billion for the 1981–85 period. Over the period the Bank has approved loans totalling $443.4 million. 56% of the total cost is for the paving of the 1 500 km BR 364 road. Settlement of new areas takes 23%, rural development 13%, land tenure services 3%, health 2% and environmental protection including Amerindian affairs 3% complete the programme. The Bank states (Botafogo, 1985) that it became involved following the government's concern that uncontrolled and spontaneous settlement could harm regional ecology. Objectives of Polonoroeste include steering continuing migration away from fragile and/or ecologically exceptional areas, including Amerindian areas, and encouraging sustainable agricultural practices which do not threaten the region's long-term potential.

Under Polonoroeste the state governments of Rondonia and Mato Grosso aim to establish agricultural colonies and improve existing settlements. The programme aims to reduce crop losses through improved storage and transport, reinforce extension and research and promote use of fertilisers, herbicides and insecticides. There are also plans to regularise and facilitate land tenure, improve education by teacher training and schools, and improve health by water supply, clinics and particularly malaria control.

Environmental and social aspects of Polonoroeste – the World Bank proposals

Conservation units

During the early stages of Polonoroeste the World Bank realised that the programme would have a significant impact on the region's forest ecosystems. At a conference held in 1982 Dr Robert Goodland, the Bank's ecologist, estimated that approximately 100 000 square kilometres could eventually be lost. At the meeting, Dr Goodland explained that, in order to manage in perpetuity some samples of forest and wildlife, a conservation system of eight units, totalling 2 000 000 hectares, would be financed by the programme.

Brazil has a variety of types of protected area each with legal protection against invasion and forest clearing. The 8 units proposed under Polonoroeste include a national park, 3 Biological Reserves and 4 Ecological Stations.

The main items of cost for this 8 unit system are demarcation and surveillance of the boundaries of these conservation units, establishment of staffed and equipped control posts, construction of administrative and residential buildings, landing strips, and aerial surveys and provision of a four-seater monitoring aircraft and vehicles.

In view of the complexity of the Amazonian ecosystems and the lack of scientific information about them, it is difficult to design management strategies for these resources. Baseline ecological studies, biotic inventories, hydrometeorological and biogeochemical studies will therefore be included in the World Bank's research programme.

Forest Reserves

Creation of a system of sustainable commercial forests in Amazonia, while planned, has yet to be achieved and will, in any event, be a complicated and drawn-out process. A forestry development team has been created for research and field studies. Some sites have been identified for forestry development. The Bank proposes that these Forest Reserves are demarcated, protected, and inventoried. Four new forest control posts are to be created, staffed, and equipped and the three existing posts be strengthened. Other project components will monitor deforestation from the government's Remote Sensing Laboratory, and seek to increase the efficiency of extraction and use of timber, including its possible use as fuel.

Tribal peoples

Information on the Amerindians in the Northwest Region is scarce and often contradictory. Over 6 500 Amerindians belonging to about 34 tribal groups are thought to live in the area. There are 58 Amerindian villages in the region; 26 in Rondonia and 32 in Mato Grosso. These groups differ greatly both in size and in degree of contact with outside cultures. Some groups own and operate tractors and may even hire non-tribal labourers, while others are uncontacted, or in only intermittant contact, with other Brazilians.

In Brazil, tribal peoples' rights to their lands are guaranteed by the constitution. The World Bank, in its tribal peoples' guidelines also recognises that the single most effective action to protect tribal peoples is to safeguard their lands. Under Polonoroeste, the Bank proposes to strengthen tribal peoples' land protection by establishing a network of Indian Reserves. Following "interdiction", Amerindian lands are marked on maps "delimitation"; then they are physically "demarcated" on the ground, usually by a 6 metre wide cleared swath, and accorded official reserve status. There are plans to accelerate the delimitation and demarcation process, resolve border disputes, evict illegal settlers from within reserves and provide surveillance to prevent renewed trespass. It is proposed to improve Amerindian health by immunization campaigns and control of malaria and TB. Clinics and health workers will be provided.

Zoning of settlement areas

Since the soils of the region are highly variable and largely unsurveyed, appropriate land use can be designated only when soil fertility and land suitability have been assessed. Therefore a 1:250 000 scale soil and land suitability survey is being carried out for all of Rondonia. The Bank plans to carry out more detailed surveys (1:50 000) on the more promising sites revealed by the broader survey. This will enable land use to be more reliably tailored to the carrying capacity of the land; the surveys will enable colonies to be more appropriately sited, and should lead to a prudent dimensioning of each land holding.

The Impact of Regional Plans: Agriculture and Industry 53

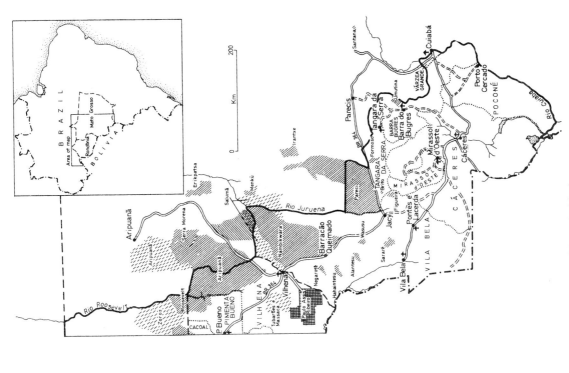

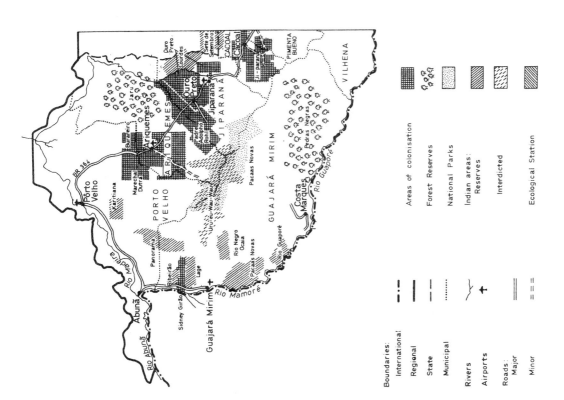

Figure 13. Map of Northwest Region Programme, Polonoroeste (Goodland, 1985). Reproduced with kind permission of Manchester University Press

5.2.2 Criticism of Polonoroeste and World Bank involvement

Since the initiation of Polonoroeste in 1981 a large grouping of Non-Governmental Organisations (NGOs) from Brazil, USA and other parts of the world have come together in a lobbying campaign against Polonoroeste and the Bank's involvement. Numerous independent sources in Brazil have provided evidence that the Programme was having huge *ecological* and *social* impact on the Northwest Region. Some of the major criticisms of NGOs are outlined below.

Criticism 1

The Polonoroeste area now has the highest rate of deforestation in Brazil's Amazon. If present trends continue, the state of Rondonia, an area the size of Britain, will be deforested within a decade. Many NGOs believe that Bank investment has accelerated deforestation in the region. Rather than being "consolidated", as was the Bank's intention, reports indicate that many settlers are abandoning their cleared land. Due to inappropriate agricultural techniques, harvest yields typically decline after the first year or so, as discussed in Chapter 4. Thus many farmers need to make new clearings every year. Then, when the whole plot is cleared, they move on again. Thus deforestation continues.

Criticism 2

In many instances settlers are selling their plot to large landowners for cattle ranching, a land use which numerous past experiences in the Amazon have shown will also be unsustainable (see Chapter 4). Jose Lutzenberger also points out that, although Polonoroeste has only been operating for a few years, already there are examples in Rondonia where entrepreneurs have bought up 6–12 plots. Thus it seems that the process of agricultural land concentration, which forced the poor off land in the northeast and south, is already beginning in Rondonia.

Criticism 3

By focusing its finance on the construction of roads and other infrastructure many critics believe the Bank has contributed to untenable migration rates into Rondonia which result in escalating deforestation etc. Over half the total finance for Polonoroeste has been spent on the BR 364 road. It is difficult to envisage how this approach could ever encourage "consolidation" of agriculture in the region.

Criticism 4

The terms of the World Bank Loan Agreement for Phase 1 of Polonoroeste have frequently been disregarded. The agreement said measures must be promptly implemented to protect the environment and tribal peoples. Many believe the Bank has lost control over, or will not take effective measures to control, the destruction taking place in the region. An example of how terms in the Loan Agreement have been broken has been given by Jose Lutzenberger. The Guapore Valley, the last large area of untouched forest in Rondonia, was about to be opened up by the BR 429 road in November 1984. Yet it flouts two conditions:

a) in one section of the Phase 1 Loan Agreement it is agreed:

> 'to discourage the agricultural exploitation of areas which have been determined to be unsuitable for agricultural development or of areas whose suitability for agricultural development has not yet been determined'.

But the official soil survey shows that many of the proposed settlement sites in the Guapore area have large areas of soil unsuitable for agriculture.

b) another term of the Agreement states:

> 'The Borrower and the Bank agree that the strengthening of the measures to protect the indigenous Amerindian population in the programme areas is essential to the successful carrying out of the project. To this end, the Borrower shall take all necessary measures to put into effect promptly the special project for protecting the interests of the Amerindian communities located in the programme area'.

Yet this proposed new road goes through an area that is almost certainly regularly hunted by *unknown* Indians who have killed colonists and rubber tappers in the area during 1981 and 1983. In such a situation it is not possible to carry out protective measures such as demarcation of Indian reserves or immunization of Amerindians.

Respected Brazilian environmentalist criticises Polonoroeste

Jose Lutzenberger, an agronomist and engineer who is one of Brazil's most respected environmentalists, criticises the whole concept of encouraging farmers to migrate to Amazonia. He believes the 'settlement schemes are conceived precisely in order not to have to face social justice in other parts of Brazil, not to carry out agrarian reform and not to change agricultural policies in other regions'. At a recent US Congress Hearing he said 'it is important to be aware that the principal social and political objective is to transfer our agricultural poor mainly from the northeast and south of Brazil, to the Amazon'. It is calculated that there are at least 2 500 000 landless poor in Brazil today. Lutzenberger says that 'Polonoroeste is designed as a safety valve for the political and social pressures caused by them. There is in fact no shortage of land in the south (of Brazil) except the shortages caused by the concentration of land holdings'. At the US Congress Hearing he asked whether the World Bank should provide funding for a project which:

> 'makes it easier and socially safer for the powerful in the northeast and south ... If the World Bank wishes to help us with our problems, why does it not invest more in projects which help fix the agricultural poor on their own lands in the south and northeast? And why does it not invest in research to improve the economy of the caboclos and rubber tappers who already live in the rainforest'.

Jose Lutzenberger. *(A. Cowell/Central ITV)*

Lutzenberger also points to the illogicality of moving migrants from the south, where the soils are generally fertile, into an area like the Amazon where it seems most soils are poor.

Criticism 5
In spite of early problems the World Bank actually accelerated lending in late 1983 for the third phase of Polonoroeste which plans to settle 15 000 more families. The purpose of this accelerated loan disbursement was to help Brazil maintain development momentum in the face of continuing world recession. Yet many critics feel it violated basic principles of sound management by accelerating funds for a programme which was already facing difficulties in managing the resources at hand.

Criticism 6
Another major criticism is that the destructive land uses resulting from Polonoroeste are actually uprooting people who are implementing sustainable land uses. Development interests often choose to forget that many rainforest areas are not in fact empty wilderness devoid of people. In Rondonia, Lutzenberger (1985) points to the sustainable land management practices of 3 groups: the Amerindians, the rubber tappers and the "caboclos" (refer to chapter 2, section 2.1 and chapter 6).

5.2.3 NGO lobbying of World Bank US Congress

In 1984 concerned NGOs raised these criticisms at a number of US Congress Hearings which had been set up to investigate the environmental record of Multilateral Development Banks such as the World Bank. A number of NGOs also wrote to the then World Bank President, A. W. Clausen. They asked him whether specific measures would be taken to correct the problems associated with Polonoroeste. In the letter to President Clausen NGOs urged:

> '... the Bank to reconsider the implications of funding programmes such as Polonoroeste. These programmes which serve as 'escape valves' for the human consequences of government economic policies and gross inequalities in land tenure in other parts of the country, which have resulted in the migration of millions of rural farmers over the past decade and a half. The financing

Cattle on ranch. *(T. Gross/Oxfam)*

of the settlement of tens of thousands of families in ecologically dubious and unsuitable areas of the Amazon is clearly not a viable solution to these complex problems'.

NGOs which signed the letter included among others: Natural Resources Defence Council (NRDC), National Audubon Society, Survival International, Brazilian Anthropological Association, Brazilian Bar Association, Centre for Science and Environment (India), Association for the Protection of Nature of Rio Grande do Sul, and Feminine Democratic Action of Rio Grande do Sul.

This letter was not answered by The Bank's President, Mr Clausen, in person. His Chief for the Brazil Division, Roberto Gonzalez Cofino wrote a brief letter which did not address any of the points raised by the NGOs. Cofino's letter was unanimously condemned. Indeed, Congressman Robert Kasten Jr, who had taken a close interest in the environmental record of the MDBs, in January 1985 wrote to President Clausen saying:

'... the response from the World Bank was at best a brush-off, but frankly, more correctly described as an insult ...As you know better than anyone else, securing support for US contributions to MDBs is difficult at best. That the World Bank would respond in such a cavalier fashion to groups and individuals who would otherwise support their programs is most difficult to understand'.

In March 1985, as a result of this type of high level representation, *the World Bank halted nearly US $250 million* of remaining disbursements on Bank loans *for Polonoroeste*. The original Bank commitment was $435 million so $185 million had been disbursed by March, 1985. This was the first time the Bank had ever halted loans for environmental reasons. Bank officials stated that no more payments for Polonoroeste would be made until the problems associated with protection of natural areas and Indian lands had been resolved. In summer 1985 Congressman Robert Kasten and interested NGOs met with President Clausen to discuss the Polonoroeste situation.

5.2.4 The World Bank's response to criticisms about Polonoroeste

The suspension of Bank funds for Polonoroeste in March 1985 indicated that the Bank admitted to at least some of the criticisms about Polonoroeste. To quote Jose Botafogo's (the Bank's vice-president for External Relations) letter to the editor of The Ecologist in autumn 1985:

'It is true that there have been problems and this is understandable given the dynamic nature of the growth and change taking place. For example, the programme has been comparatively more successful with infrastructure development than with institution building or services to farmers. The special project for protection of Amerindians, involving, among other things, full establishment of five reserve areas, has not moved as well as planned.

As a result of these and other implementation difficulties, the Government of Brazil earlier this year took the initiative to have disbursements from outstanding Bank loans supporting the programme held in abeyance until a *remedial action plan* could be discussed and agreed with the Bank. This was done and progress is now being made'.

In the same letter Jose Botafogo responded to criticism that Bank involvement in Polonoroeste had accelerated environmental destruction and jeopardised Amerindian people in the region. He stressed that the Bank became involved in order to minimise uncontrolled migration into Rondonia *which was already occurring*. He said the Bank also aimed to:

i) steer continuing migration away from Amerindian areas and fragile and/or ecologically exceptional areas;
ii) to encourage sound agricultural practices.

Significantly, at the time of writing he did not claim there has been any on-the-ground implementation of these objectives. The section of his letter referring to Polonoroeste ends:

'In retrospect, the easiest course for the Bank might have been not to get involved at all in the Polonoroeste programme, to "play it safe" and thereby possibly avoid public criticism. This would not, however, have prevented environmental problems and jeopardy to tribal people from occurring as a result of continued uncontrolled settlement'.

In addition, the World Bank might stress that, despite problems at implementing its own environmental and tribal peoples' safeguards in projects like Polonoroeste, its record is far better than many other Third World investors. This is a very important and valid point. Critics need to bear this in mind. It would be disastrous if they carried out "over-kill" on MDBs whilst forgetting about the environmental and social record of even bigger investors in the Third World eg. private banks and multi-national corporations. Few people doubt that World Bank environmental policies and procedures are way ahead of most other such investors. The issue at stake is whether or not World Bank environmental policies and procedures can actually be implemented.

By January 1986, the World Bank and Brazilian government had prepared and agreed a Remedial Action Plan for Polonoroeste. In the same month funding for the programme started up again. Simultaneously, NGOs in North America and Europe, including NRDC, Oxfam, Friends of the Earth and WWF, set up a group to look at the environmental track record of MDBs in general, as well as bilateral aid agencies, such as the British government's Overseas Development Administration. NGOs continue to monitor the Polonoroeste situation. New investigations into the social and environmental impact of other large development schemes, including the new Inter-American Development Bank project in Acre, are now also underway.

Activities linked to this chapter

Your teacher has details for the activities listed below:

* 'Filming in Rondonia', preparing a film storyboard (Activity 10).
* 'Sorting out opinions', an activity examining the beliefs of some prominent Brazilians concerning development in the Amazon (Activity 11).
* Analysing the TV film 'Banking on Disaster' (Activity 12).

Chapter 6

Towards Balanced Development

6.1 The Amazon Pact and the Brazilian Government's Forest Policy Report (after Johnson, 1985)

Since the late 1970s there have been a number of important developments with regard to Amazonian development policy. Two particularly hopeful initiatives have been The Amazon Pact and the Brazilian Government's Forest Policy Report which are outlined below. Another important regional initiative, the rubber tappers "Extractive Reserves" proposal, is considered in the next section.

The Amazon Pact (Treaty for Amazonian Cooperation) was signed on 3rd July 1978 by Bolivia, Brazil, Colombia, Ecuador, Guyana, Peru, Surinam and Venezuela. This Treaty was "inspired by the common aim of pooling benefits ... so as to raise the standard of living of (the signatories') peoples ... and to achieve total incorporation of their Amazonian territories into their respective economies."

Clearly the Treaty is a pact for development cooperation. Nevertheless it makes frequent and positive reference to conservation among "responsibilities inherent in the sovereignty of each state". In fact Article VII of the Treaty refers to "the need for the exploitation of the flora and fauna of the Amazon to be rationally planned so as to maintain the ecological balance in the region and preserve the species".

Meanwhile Article XI commits the contracting parties to joint studies and measures in order to increase the rational utilization of the human and natural resources of their respective Amazon territories, and Article XIV commits the parties to "cooperate in ensuring that measures adopted for the conservation of ethnological and archaeological wealth of the Amazon region are effective."

The Treaty established Permanent National Commissions to carry out the decisions of an Amazon Cooperation Council which was to meet annually in a national capital of a member state, rotating alphabetically.

Taken as a whole, the Pact represents, from the environmental point of view, a moderately hopeful departure. Moreover it offers the best available forum for a major regional strategic initiative promoting environmentally sound development in the Amazon basin.

The second initiative, the Forest Policy Report, occurred at the national level. It took the form of a report by an official, high-level, interministerial commission of the Brazilian government. This Commission began its work in 1979–80 in the wake of strong domestic environmental protest at the Government's 1978 proposals to grant unrestricted timber rights to multinational logging companies over large areas of rainforest. These 1978 proposals were shelved in the face of protest, especially from local cattle ranching and timber interests. Meanwhile the Commission was formed to draw up a comprehensive plan for the management and development of the forested region of the Brazilian Amazon basin.

This report was completed by 1982 and has since been sitting, unpublished, on the desk of the President. Though the public remains unaware of its contents, it has been the focus of considerable controversy. According to one of its authors, Professor Mauro Reis, chief of the National Forest Development Institute, the report set aside great national forests, Indian reserves, and reserves for scientific study as well as indicating regions for appropriate types of development.

The report was drafted by a group of government officials and academics. It was the first wholesale attempt to produce any sort of forest lands policy for the Brazilian Amazon as a whole. The report contains firm guidelines on how land development should take place, and argues the need to preserve much of the Amazon rainforest, while speaking of the danger of unrestricted development. This was, apparently, the first time that these issues had been formally raised and analysed at governmental level. The report drew boundaries marking areas that could be logged and those that might be used for agriculture. Existing Indian reserves were expanded by the report and a system of national parks set up.

The World Conservation Strategy

The World Conservation Strategy (WCS) was commissioned by the United Nations Environment Programme (UNEP). Prepared by the International Union for Conservation of Nature and Natural Resources (IUCN), it attempts to represent a consensus on conservation policy in the context of world development. The aim of the WCS is to help advance the achievement of **sustainable development** through the conservation of living resources. The Strategy:-

1 explains the contribution of living resource conservation to human survival and to sustainable development;
2 identifies the priority conservation issues and the main requirements for dealing with them;
3 proposes effective ways for achieving the Strategy's aim.

The Strategy is intended to stimulate a more focused approach to renewable resource conservation and to provide policy guidance on how this can be carried out. It concentrates on the main problems directly affecting the achievement of conservation objectives; and on how to deal with them through conservation. In particular, the WCS identifies the action needed both to improve conservation efficiency and to integrate conservation and development.

Key sections of the WCS outline a framework for national and regional strategies and emphasise that, in the long term, individual conservation initiatives cannot succeed in isolation. As the diagram here demonstrates, they must be planned and implemented as part of an overall development strategy.

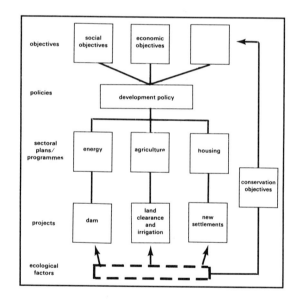

The need to integrate conservation with development: an example.

Attempts to minimize the ecological harm (and hence the social and economic harm) of a dam rarely succeed if ecological factors are considered only at the project stage. By then the dam is a key component of other major projects (like land clearance, irrigation, and new settlements), themselves essential parts of several sectoral programmes. These programmes are often expressions of social and economic policies from which ecological considerations are entirely absent. Unless ecological considerations influence the development process along with social and economic considerations - and unless there is also an explicit policy to achieve conservation objectives - the prospects of avoiding ecological harm and of making the best use of living resources are dim. Thus when ecological factors are considered only at the point shown at the bottom of this picture their influence is usually limited or negative. Instead, for development policy to be ecologically as well as economically and socially sound, the empty circle at the top of the picture needs filling as shown.

The authors of the report had hoped that this policy would go to Congress and be the subject of a great national debate. This has not, however, taken place. The reason why the forest policy document was shelved, according to Paulo Yokota, head of the government's land settlement agency, was that it was seen by other parts of the government, and certainly the business establishment, to be the work of radical environmentalists. The document's stress on forest protection has earned it the criticism that its authors talk about trees as the most important asset of the region. It has become the standard counter argument to contrast the importance of trees to that of people.

Thus at present there is confusion and confrontation. As Johnson (1985) shows, what is required is an overall strategy based on inter-disciplinary multi-resource analysis. It is precisely such analysis which is called for by the World Conservation Strategy (IUCN et. al. 1980) with its requirement for national and regional conservation/development strategies. Since environmentalists claim to see things as a whole and recommend that governments themselves start holistic planning, it is essential that the environmental interest should attempt an over-all perspective.

6.2 Rubber tapper "Extractive Reserves": a new conservation – development initiative for Amazon forests

An interesting conservation initiative is gathering momentum in Brazilian rainforests and the impetus is coming from Amazonia's rubber tapper communities.

Put briefly, Brazilian rubber tappers are calling on their government to implement new and far-reaching measures to integrate the conservation of the country's rainforests with the needs and chosen lifestyle of communities such as themselves, communities which have been living in the rainforests for generations.

Mary Helena Allegretti and Stephen Schwartzman (1986) have documented the development of the new rubber tapper movement. At an important meeting held in Brasilia in October 1985, representatives of the rubber tapper communities strongly protested to the

Rubber tapper in forest. *(A. Cowell/Central ITV)*

government that, to date, they had been excluded from Amazonian development schemes. In the state of Rondonia, for example, many rubber tappers have been expelled from their forest lands to make way for migrants from south and north-east Brazil who are being attracted into the region by the roads and agricultural settlement colonies which continue to be funded by government and financial institutions such as the World Bank.

The rubber tappers persuasively argue that such development schemes are totally inappropriate, not only because they fail to meet the needs of their own communities, but because they cause massive rainforest destruction and typically result in agricultural schemes where crop yields cannot be sustained for more than 2 or 3 years, due to soil erosion, weed invasion etc. So the rubber tappers want a major rethink for Amazonian development policy. They are demanding that, in future, their communities should be involved in the identification and implementation of all development projects and that these include stringent provisions for the conservation of the rainforests which rubber tappers use and inhabit. The rubber tappers want the government to acknowledge fully the economic importance and environmentally-sound basis of their way of life and to provide backing for the establishment and development of "extractive reserves". These would be rainforest areas owned and sustainably managed by rubber tappers for latex, Brazil nuts and a range of other forest products.

Rainforests

Their proposal is based on the first-hand experience of development fiascos in Amazonia. Many of the rubber tappers are descendants of the northeasterners who migrated into Amazonia during the rubber boom at the turn of the century. This was the period of peak Brazilian rubber production, when Brazil had a monopoly. Profits were huge and conditions were brutal. "Rubber barons" demanded labour in payment for debts, a system which still exists in many areas and is called "captivity" or "slavery" by the rubber tappers.

When Malayan plantation rubber became widely available on the world market, the boom collapsed. Some of the rubber tappers stayed on in the rainforest, planting small agricultural plots, hunting, fishing and selling rubber to small merchants who travelled the rivers. In the mid-1960s, agri-business entered Amazonia, cattle ranching began and price supports for Brazilian rubber ceased. Many "rubber barons" moved out and this favoured the emergence of "free" or autonomous rubber tappers – families not

---

Rubbertappers' leader gunned down in his backyard
Fighter for Amazon ecology murdered

Jan Rocha in Sao Paulo

THE leader of Brazil's rubbertappers and internationally known defender of the Amazon rain forest, 44-year-old Francisco Mendes, has been assassinated at his home in Xapuri, in the state of Acre.

He died instantly after being shot in the chest as he went out into his backyard. Two men were seen running away from an empty house next door, but local police failed to capture them.

Chico Mendes, as he was known, had received many death threats because of his fight to save the forest from destruction. He was president of the Rural Workers' Union in Xapuri, leader of the National Rubbertappers' Association, and founder of the Union of Forest Peoples, an alliance of Indians and rubbertappers.

Mendes led the rubbertappers, whose existence depends on the survival of the forest, in *empates*, the physical blockade of machines sent in by land-owners to clear the tree for cattle farms.

In Washington last year, at a prizegiving ceremony, he denounced the involvement of large American companies like Xerox and Georgia Pacific, the Dutch company Bruynzeel and the Japanese Toyomenka in rain forest devastation:

"By buying expensive mahogany furniture, you Americans are helping to finance the destruction of the last forest reserves on the planet," he told his audience.

He successfully urged the World Bank and the Inter American Regional Bank not to finance the highway planned to link Acre's capital, Branco, to the rest of Brazil, until serious environmental protection measures had been taken.

These activities earned him the hatred of those in Acre for whom roads, sawmills and cattle farms mean progress and above all profit. For them he was no more than an agitator.

Several attempts had already been made on Chico Mendes's life. A few months ago the military police warned him that a gunman had been hired to kill him and provided him with a bodyguard. He never announced his movements in advance.

Union leaders in Acre suspect two brothers, who are members of the right-wing landowners organisation UDR, of involvement in the killing.

In a recent telex to the President of the Republic, the Minister of Justice and the head of federal police, they and other organisations denounced the two brothers as gunmen who were wanted by the courts in the south of Brazil.

The body of Mendes was taken to the capital for a post-mortem and embalming. It will be flown back to Xapuri for the funeral at the weekend which will be attended by national leaders of the Workers' Party, to which he belonged, and of environmental and human rights organisations.

Thousands of rubbertappers are also said to be on their way by river and road for the funeral.

Most of them, like Chico Mendes, are the sons of men who were brought to the Amazon during the Second World War as part of an allied plan sponsored by the US Government to boost Brazilian rubber production after the fall of Malaya.

Known as the rubber army, they were abandoned in the jungle by the authorities, and many died.

Mendes had been honoured with several international awards. Last year he received the Global 500 prize of the United Nations Environment Programme, awarded to the 500 people deemed to have done most for the environment during the year.

'The Guardian', December 24 1988

indebted to local "rubber barons" and who frequently had title to the forest land they used to extract latex.

It is these autonomous rubber tappers who are spearheading the present demands for "extractive reserves". There are hopes that their representations have not fallen on stony ground. There has been an extremely positive response from the Brazilian government, media and organisations such as the World Bank. As a result of their public presence, and the fact that the rubber tapper leaders have quickly identified the pressure points in national and international bureaucracies, the "extractive reserve" concept has undoubtedly been given official life. The Brazilian government has agreed that there must be no further encroachment on rubber and Brazil nut producing areas by dividing rainforest areas into lots, as in the typical model of tropical forest agricultural colonisation schemes. The state governments of Acre and Rondonia have now identified forest areas as extractive reserves and the World Bank has officially endorsed the proposal, describing this Amazonia conservation and development initiative as "the most promising alternative to land clearing and colonisation schemes, which are often questionable in environmental terms". Of course, it remains to be seen whether these words of support will be translated into real, on-the-ground support for the people and forests of Brazil's Amazonia.

The position in October 1988 is that 12 reserves covering 2 000 000 hectares have been established in 5 Amazonian states. However, there are concerns that the crucial element of local control is already lacking in some of these reserves. The 2 reserves in Acre seem to be working well under community management but in other states rubber tappers still doubt whether government and Multi-lateral Development Banks are genuinely committed to the original extractive reserve concept.

6.3 Identifying and implementing an "appropriate" Amazonian development strategy

6.3.1 An introduction

Much of this book has focused on the extent and rate of deforestation in Amazonia and the issue of the sustainability/desirability of one of the region's most ambitious development programmes, Polonoroeste. Although individual opinions about the precise deforestation situation and schemes like Polonoroeste will vary, most people agree that, to succeed, any future Amazonian development strategy must start with a careful consideration of *"appropriate" conservation and social/economic objectives* for the region. Refer to the Box in section 6.1. Arguably, identifying the former is less controversial than the latter. Many countries around the world at least pay lip service to the three key conservation objectives put forward in the World Conservation Strategy. Of course, even a superficial glance at the wide range of development strategies/policies now being implemented by different governments around the world makes it abundantly clear that getting agreement on "appropriate" social/economic objectives is far more difficult.

This section examines just one paper, by Dr Philip Fearnside of INPA, which tries to *evaluate* and *rank* a range of development options for Amazonian **terre firme** (unflooded upland) rainforest. Fearnside (1983) begins by setting out what *he personally feels* are the nine "appropriate" objectives for an Amazonian development strategy/policy. When studying this section, which is a resumé of Fearnside's paper, it is important for you to bear in mind that other individuals and many governments would put quite different emphasis on some of Fearnside's objectives. Remember, development strategy objectives can be, and are, hotly debated.

6.3.2 Evaluating and ranking development options for Amazonian "terre firme" rainforest (based on Fearnside, 1983)

Fearnside (1983) begins by saying that any planning must start with a careful consideration of what development objectives are appropriate. The frame of reference is crucial and Fearnside believes that several conditions must be met if the following objectives are to be reached.

The conditions include:
- the maintenance of human population below carrying capacity;
- a relatively equal distribution of agricultural land in the region;
- a limit to total consumption, presumably including a limit to maximum consumption per capita.

Fearnside (1983) next states what he believes are the 9 appropriate objectives for an Amazonian development strategy/policy:

Objective 1: Agricultural sustainability
Agricultural sustainability requires a reasonable balance of nutrients in the system, with compensation needed, e.g. via fertilisers, for losses from leaching, soil erosion and nutrient export in harvested agricultural products. Other requirements for continued productivity, such as prevention of soil compaction, must also be met. Energy sources should be renewable. Probabilities of destruction by pests or diseases must be low.

Objective 2: Social sustainability
Social sustainability requires that a particular land use system be maintained, in practice, by society. To survive, a system must remain profitable over time. Sustainability can be threatened by fluctuation in product yields, e.g. due to pests, and variations in market prices for the products, or in prices and availability of inputs such as fertilisers. The regulations required for the system's functioning must be enforceable. Whether or not people accept the social assumptions underlying a particular land use system can affect its long-term sustainability. For example, is highly unequal distribution of income and/or land acceptable to the majority of people?

Objective 3: Unsubsidized economic competitiveness
Agricultural activities must be attractive without reliance on economic subsidies from outside. Distortions introduced by such subsidies as tax incentives and very low interest loans have a way of becoming self-perpetuating, even when the land use system clearly does not deserve government support.

Objective 4: Maximum self-sufficiency
Dependence on imports of energy supplies, other agricultural inputs and basic foods is likely to lead to long-term risks from price variation and supply interruptions. Self-sufficiency must not be confused with isolation from trade, necessary, within limits, for all agroecosystems that might be planned. Some cash crops are needed to supply funds for purchase of goods which cannot be locally produced; the challenge is to prevent loss of self-sufficiency in locally producible products.

Objective 5: Fulfilment of social goals
"Minimum living standards", as measured by various criteria, e.g. access to clean water, medical services, education, must be available to those supported by a particular development option. For each criterion, the probability of an individual or family failing to meet the standard must be kept as low as possible. Employment generated by different development options must also be considered, as must an option's effect on the fair distribution of income. On a regional scale, a land use option's installation costs can also be important. For example, opting for an expensive development scheme could mean that social goals are not fulfilled elsewhere in the region.

Objective 6: Consistency with maintenance of areas for other uses
A development strategy must ensure that adequate areas are available for ecological, Amerindian and other types of reserves requiring undisturbed rainforest. Reserve boundaries, once created, must be respected. Land uses surrounding such areas must not create pressures which lead to destruction or damage of reserves, e.g. due to land invasion.

Objective 7: Retention of development options
Land uses should be avoided if they prevent other possible future uses, should a change be necessary. For example, a pest/disease problem could force the abandonment of a particular agricultural use.

Objective 8: Minimal effects on other resources

Both social and biological effects, from development activities on an area of land, can damage other areas or resources e.g. pollution can affect nearby areas. Some development options lead to uncontrolled expansion of human activities on neighbouring resources.

Objective 9: Minimal macro-ecological effects
The loss of species diversity (genetic resources) is seldom considered. Nor are the risks of regional/global climatic changes due to specific development options considered by most planners. Long term costs of ignoring these potential problems could be very high.

6.3.3 Comparison of development options

With the objectives discussed above in mind, as well as the potential for conflicts among objectives, it is useful to compare some of the development options used and proposed in the Brazilian terra firme (unflooded upland) rainforest. No single option should be considered desirable for the entire Amazon, but rather a patchwork of areas in different uses. Philip Fearnside emphasises that all possible options have drawbacks, although some are clearly more desirable than others. He believes that the idea that a single amazing development option exists only awaiting discovery by researchers is a myth. Although research can indeed provide better agroecosystems, and should be supported and pursued at many times its present rate, the faith that research results will someday overcome any given agricultural and environmental limitations is misplaced. Such faith can and does lead planners to dismiss concern for future consequences of present development decisions.

Below, 9 possible land uses are compared using the criteria outlined above.

1 Untouched Forest
Untouched Amazonian rainforest is clearly a sustainable use over any time horizon for human planning. Provided forest is indeed untouched, the only threat to its sustainability would either come from region-wide climatic changes or from isolation into patches that are too small in the long term for some plants and animals to maintain themselves in, e.g. predators with large territory requirements.

Social sustainability is a problem. Untouched forest is vulnerable to conversion because many developers see it as providing unacceptably low rates of economic return compared to other land uses. Presently there is a lack of social controls to prevent human disturbances and many changes resulting from disturbances are irreversible, e.g. species extinctions.

In the short term, untouched forest is not competitive with other uses. An exception is the value of forested land as a commodity for speculation. But here expectation of future conversion to other uses, say mining, forms the theoretical basis of the rise in value, even if each landowner does not intend to undertake the conversion himself. It is difficult to assess long term competitiveness. However, long-term holding of large forest tracts for future economic exploitation could definitely be a wise use in economic terms. The value of wood alone in the Brazilian Amazon has been estimated at over US$1 trillion at current international prices for hardwoods on commercial markets (United Nations data). Note that as well as reserving areas for *future exploitation*, sufficient areas should be *preserved untouched* in order to maintain genetic diversity with guarantees that human disturbance will not be permitted at some future date.

Social goals are served by untouched forests from very poorly to very well, depending on the goals in question. Employment generation is low, the main jobs are as forest guards and research workers. Living standards are not maintained from forest products extraction in the short term, and so depend on the scale of salaries paid by whatever governmental or private organization would be responsible for the tracts. Low installation cost is a point in favour of "untouched forest", since both public and private funds are freed for other social requirements such as education and health. These requirements also include the installation of adequately funded sustainable agroecosystems elsewhere, either in Amazonia or elsewhere in Brazil. Ideally, these sustainable agroecosystems will probably be located on the varzeas (seasonally flooded forests) of Amazonia.

Whether social goals are fulfilled by keeping rainforest in the "untouched" category depends on whose social goals are in question. Excluding poor squatters from untouched forests which have been set aside for the future use of government or private owners clearly puts these individuals/groups in conflict. Should Amerindians be permitted to live in "untouched forest"?

Untouched rainforest has the advantage of not adversely affecting adjacent areas. Obviously there is no human encroachment from untouched forest to other areas. Other positive points are the retention of development options, minimal adverse effects on other resources and the absence of undesirable macro-ecological effects.

2 Forest products extraction

Extraction of forest products can provide income from rainforest areas in ways which, for many products such as Brazil nuts, can be sustained indefinitely from an agricultural point of view. Products removed represent a relatively small drain on nutrients over a large area due to the small quantities involved, although the nutrients often are in concentrated form, e.g. nuts. Most important, forest nutrient cycling mechanisms are left unharmed.

Income from extractive production can be significant: the USA alone imported US$ 16 million worth of Brazil nuts in 1977. However, short term competitiveness is typically low, because incomes from many forest products extraction operations are currently low on a per-hectare basis. Of course, future increases in product prices and in the range of products could easily change this.

Social sustainability is likely to be low for many forms of forest extraction. Game hunting is unlikely to provide a continued production, except in isolated areas, because most regulations intended to conserve game populations are unenforceable, e.g. due to lack of forest guards. Social sustainability is often unlikely due to competition with other land uses. Rubber tappers in Acre, for example, have been violently expelled by ranchers or land speculators. On balance it is probably best to evaluate social sustainability as "unknown".

Rubber tapper communities form a majority of the population in some Amazonian states. Many rubber tappers are still among the poorest in Amazonia and when rubber extraction from rainforests is still controlled by "rubber barons" it is hard to accept that rubber tapper self-sufficiency and fulfilment of social goals is being achieved. However, the independent rubber tapper communities now emerging in some areas could, in future, significantly change the situation. Extraction of forest products does not conflict with land uses in adjacent areas and it leaves options for other uses almost entirely open. Damage to other resources and macro-ecological effects are also minimal.

3 Agroforestry

There is now considerable interest in variations of "taungya", a particular type of **agroforestry** with a long history of use in SE Asia. It involves planting annual crops with commercially valuable tree species like teak. The trees take the place of "bush fallow" in the shifting cultivation cycle, giving a valuable yield at the end of each cycle. Of course the trees, which are felled and carried away, do not provide the ash of burned secondary vegetation as in unmodified shifting cultivation.

Prospects for agricultural sustainability are moderate, but fertiliser inputs would probably be needed after a few agroforestry cycles. The same potential biological and other problems applying to forestry must be considered for agroforestry.

Social sustainability of taungya should be quite high, since the system produces both food and cash crops. Also the tree crop formed during the cycle would be of sufficient value to justify defence against land invasion or other pressures to switch to other land uses. However in Amazonia there would be risks of taungya system abandonment after the first valuable tree harvest. Unlike in southeast Asia virtually none of the pioneer settlers in Amazonia have a tradition of allowing "bush fallow" to regrow on their plots.

Agroforestry's competitiveness in the short term should be similar to that of annual crops, only moderately attractive given today's product/fertiliser prices. As with other land uses involving annual crops, agroforestry systems have the advantage of producing a return beginning in the first year. The extra costs of planting trees would make agroforestry less attractive to many farmers because their trees' "future returns" are such a long way off. In the long term, once the first tree harvest is produced, the agroforestry system should be much more attractive.

A high degree of self-sufficiency is one of the main advantages of agroforestry. The annual crops produced each year can reduce dependence on outside suppliers of basic food crops. However, research into modified and more diverse agroforestry systems for Amazonia is essential. Tree species providing additional subsistence crops such as fruits and nuts, need to be encouraged, for example coconut, Brazil nut, breadfruit and jackfruit trees.

Social goals are typically well served by agroforestry. The system is labour-intensive, creating more jobs per unit area than many other options. Average income is probably very high, especially after the tree crop(s) begins to produce a return. The system is well adapted to using family labour. Income from agroforestry should be fairly stable but fluctuations in product prices will affect this as with other crops.

The effect of agroforestry on nearby reserves or managed rainforest areas should be moderate. The high investment required to establish the tree crop(s) makes spillover into adjacent areas less likely than for some other land uses. However, the annual crop phase, without the accompanying tree crop phase, could overflow into surrounding areas if individual farmer properties are too small. Perhaps more importantly, the relatively concentrated population associated with agro-forestry would be tempted to enter and hunt in adjacent forested areas.

As with other land uses involving clear-felling, agroforestry closes all options requiring rainforest cover. As agroforestry systems involve the unalterable destruction of rainforest there are doubts about recommending them for large areas presently under rainforest cover. However the systems have many clear advantages in Amazonian areas already deforested.

Agroforestry's effects on other resources can be considered moderate. Macro-ecological effects are reduced by the presence of tree cover during most of the cycle.

4 Annuals in shifting cultivation
The agricultural sustainability of **annual** crops in shifting cultivation varies from good to bad depending on whether the appropriate "bush fallow" is carried out.

Social sustainability of shifting cultivation can be either good or poor depending on circumstances. There is often a temptation to hasten the cycle or to switch to other land uses because of the low returns from shifting cultivation. Many of the pioneer smallholders lack the cultural traditions necessary to sustain the unmodified shifting cultivation system. In addition, fallow areas would be classed as "unused" or "abandoned" by many outsiders and would therefore be vulnerable to squatters.

Shifting cultivation is moderately competitive with other uses on a short term basis. Required cash inputs are low and crop returns begin in the first year. Over the long term the system's competitiveness would be reduced as the area occupied by fallows increases, and as possible options for slow-yielding crops appear more attractive. Most settlers in government colonization projects view annual crops only as *temporary* cash resources whilst establishing cattle pasture or perennial crops.

Shifting cultivation has the advantage of self-sufficiency, the farming family getting most daily needs from their own fields.

Social goals can be well served by shifting cultivation, with the exception of the farmers' extremely low incomes. Low mean income, as well as the low contribution of saleable items to the wider economy is the main reason presented by most of today's agricultural planners when condemning shifting cultivation. Fluctuation in yields can cause crop shortages but crop diversification, for example, can reduce the impact of crop failure. Shifting cultivation does provide large amounts of employment, uses family labour, requires small inputs of cash from public/private sources and helps promote even income distribution.

A major problem with shifting cultivation is that it

frequently puts pressure on adjacent areas under different management regimes. If and when pressures increase on the land under shifting cultivation, it is difficult to prevent farmers moving into adjacent forest reserves or managed forest areas.

Shifting cultivation closes options for all land uses requiring rainforest cover. Effects on other resources can occur, e.g. siltation problems arising from soil erosion, fires from burning fields spreading into forest areas. As with other uses involving clear-felling, shifting cultivation has macro-ecological effects but these are small. However, "bush fallow" regrowth on parts of the land does moderate some of the effects.

5 Annuals in continuous cultivation
The agricultural sustainability of continuous **annual** crop cultivation in Amazonian "terre firme" is not known at present. Experiments in the Peruvian Amazon have shown the need for supplying an increasing number of nutrients as fertilisers, as cultivation continues for crops such as maize, rice and sorghum. The experimental plots indicate that, at present prices, fertiliser costs can be covered by crop incomes. Results are encouraging so far. However, the requirement of continuous input of technical information and acceptance by farmers, plus the need for cash to pay for fertilisers, means the scheme is unlikely to be widely used in most of Amazonia at present.

Appropriate crop rotation schemes, interplantings of different crops, mulching, and other techniques in addition to fertilization may eventually yield a sustainable and economically competitive system. So far such a system has not been developed for Amazonia. One attempt to develop a sustainable subsistence gardening system for tropical rainforest areas in lowland Mexico uses "modules" of garden plots in association with animal raising and fish farming. Many biological factors, in addition to nutrient supply problems, need to be solved if large scale agricultural operations with annual crops are to succeed on a sustainable basis. Weeds are one such problem.

Social sustainability of continuous annual crop cultivation should be excellent, if agricultural problems can be solved. Such a system would not leave areas in "bush fallow" or low-intensity forest uses which could be classed as "idle" and open for occupation. The social tensions frequently associated with large plantations could be avoided if the continuous cultivation schemes in use included large numbers of small farmers.

At the moment, annual crops in continuous cultivation are not competitive with other land uses, except vegetables near large city markets, e.g. Belem and Manaus. In future this option may improve as better techniques are developed.

Self-sufficiency can be high or low under annual cropping schemes. Diverse food crop cultivation clearly promotes self-sufficiency, whilst large plantations of cash crops, say cotton, lead to increased dependence on the market for subsistence requirements.

Social goals are likewise highly dependent on the type of crop/system adopted. There is ample evidence that some large plantations exploit hired labourers. Obviously this is unacceptable to planners if an adequate minimum wage and reasonably equitable income distribution are stated social goals. Fluctuations in market prices and crop yields would also have varying effects depending on the crops, and the diversity of crops chosen.

Continuous cultivation of annuals should pose somewhat less threat to neighbouring areas under forest land uses than does shifting cultivation. However, concentration of population near forested areas would inevitably lead to some disturbance. Continuous cultivation systems close all other development options requiring rainforest cover.

Potential for effect on other resources through pollution is much higher than for other land uses. In addition to fertilisers, continual inputs of pesticides are needed to combat weeds and insects. Erosion and siltation are also potential problems in watershed areas. Serious macro-ecological effects can be expected from large areas under continuous annual crops because all tree cover is removed.

6 Pasture with fertiliser
The agricultural sustainability of pasture with fertiliser applications is under intensive research in Brazil's

Amazonia. Fertiliser applications on degraded pasture have produced encouraging results which government scientists believe are sustainable. Of course it must be noted that fertilization to replace lost nutrients is only one of several conditions required for long-term pasture production. For example, soil compaction and weed invasion need to be controlled.

Social sustainability of pasture with fertiliser is not at all likely at present. Even with generous government subsidies for pasture fertilisers, *very few cattle ranchers in Amazonia presently use fertilisers of any kind*. Most ranchers have ulterior motives for maximising area of land cleared rather than concentrating their financial resources on raising productivity of limited areas of pasture.

Without government subsidy, competitiveness of fertilised pasture with other land uses is doubtful both in the short and long term. Principal problems are higher fertiliser prices in Amazonia, mainly due to transport costs, and the need for more fertiliser to achieve a given effect on typical rainforest soil as compared with other agricultural regions.

Cattle ranching operations are rarely self-sufficient in basic foodstuffs since those crops do not form a part of the production system. Most social goals are poorly served by ranching. Ranching operations are usually large, often mirroring the social inequalities of northeast Brazil. Little employment is created by ranches once initial forest felling is completed. A survey on 11 ranches calculated that one job per 778 hectares of ranchland was typical.

Ranches have often posed threats to nearby forested lands. Expansion of clearings into adjacent land is a way of gaining possession, the legal formalities of title being worked out later in accord with the fact of occupation. Expansion of ranching into a number of Amerindian reserves has occurred in this way.

Ranching closes all options for land uses involving forest cover, as well as making it difficult to change to other alternative land uses without a period of fallow. Changes of land uses are difficult primarily due to aggressive competition of pasture grasses and weeds in fields planted with new crops. Aerial sowing of pasture is now widespread; on occasion aerial "bombing" of Amerindian and squatter annual crop areas with pasture grass seed has occurred. It seems that ranchers know of the aggressive nature of the grass. Of course, on these occasions the aim is to expel the Amerindians and squatters by ruining their croplands.

Macro-ecological effects are a major concern with pasture development, especially due to the large areas required by this land use. Climatological effects are likely to be most pronounced under pasture since the land is maintained free of tree cover.

7 Pasture without fertiliser

This land use is rapidly occupying most cleared areas of Brazil's Amazonia. Agricultural sustainability of pasture without fertiliser is limited to a few years. Soil phosphorus depletion, soil compaction and weed invasion soon result in declining productivity. Without government subsidies competitiveness with other land uses is poor due to low grass and beef yields. The bad record of pasture in meeting self-sufficiency/social goals, in permitting other land uses in adjacent areas, in retention of options, effects on other resources and macro-ecological effects are all equivalent in the cases of pasture with or without fertilisers.

8 "Highgrading" with replanting

"Highgrading" is a forestry system whereby only the most valuable tree species are harvested whilst leaving the remainder of the forest intact. It has long been the preferred timber extraction practice in Amazonia. In this system trees removed are replaced with planted seedlings. It appears to form the basis of sustainable forest management in Brazil's rainforests. Prospects for sustainability are good *provided* the replacement seedlings do establish themselves and grow to maturity for the next harvest. At this stage it is too early to be sure that "highgrading" with replanting will be viable on a commercial scale. Nutrients removed from the forest in the form of logs would be small relative to many other land uses. However, eventual nutrient depletions would need to be replaced using fertilisers. Reports from southeast Asia also highlight the difficulty of extracting valuable trees without damaging others when using large mechanised equipment.

Social sustainability depends on regulations for

laborious replanting operations being enforced over many years. It is difficult to predict whether the necessary replantings would be carried out in practice.

"Highgrading" with replanting is probably only moderately competitive in the short term due to the high cost of replanting and maintaining the areas around seedlings free of weeds. Future price increases for tropical hardwoods, as the world's rainforests are destroyed, could change this, giving this land use good prospects for long-term competitiveness.

As with shelterwood forestry, this system does not promote self-sufficiency. Forestry management schemes such as this require salaried labour under the direction of an appropriate forestry institution. Whether minimum wage levels and distribution of income are achieved depends on the nature and management of the forestry institution.

"Highgrading" with replanting scores high marks for consistency with other land uses, retention of development options, effects on other resources and avoidance of macro-ecological effects. However, it should be pointed out that in southeast Asia it has been found that the extraction, even of small numbers of trees, causes considerable physical damage to the forest. Logging companies usually have no incentive to minimise this damage and this greatly limits the ecological viability of such schemes.

9 Clearcutting without replanting
On an extensive scale, this development option has very poor prospects as a sustainable land use in rainforests. The risk of poor forest regeneration is high because increased light intensity at ground level favours rapid weed invasion and provides poor growth conditions for many rainforest tree seedlings. Extensive clearcutting destroys forest nutrient cycling mechanisms.

If commercial exploitation of regrowth following clearcutting was attempted, social sustainability would probably be low. In contrast to, say, a plantation scheme, secondary growth on a clearcut area would be likely to be considered by outside parties as "unused" or "abandoned". Such an area would thus be likely to be re-cleared for a more intensive use.

Clearcutting without replanting should give an attractive short-term return. The "windfall" income from clearcutting should compare favourably with exploitation costs. Most of this cost is to remove the saleable timber, establishment and growth of the forest having come to the exploiter at virtually no cost. Due to its unsustainability, the long-term competitiveness of clearcutting without replanting is undoubtedly very low.

Self-sufficiency of this development option is low and like other forestry development options attainment of social goals will vary. The risks to adjacent areas from clearcutting without replanting are probably high. Firms might be easily tempted to expand clearings into neighbouring reserves because options after clearcutting are so limited. Retention of development options requiring rainforest cover are por and erosion/soil compaction on clearcut sites lowers the area's potential for other land uses.

Other resources are only moderately affected. Soil erosion from clearcut areas on watersheds could cause siltation problems for dams. Eroded soil from hillsides is deposited as silt in streams and rivers, later "settling out" behind dams downstream. Macro-ecological effects are likely to be intermediate. Second growth after clearcutting makes some effects more temporary than in the case of land uses where tree cover is suppressed for long periods.

6.3.4 Conclusions

Philip Fearnside (1983) reaches the following conclusions.

1. Development objectives in the Brazilian Amazon need to be carefully thought through, with reference to the long-term benefit of the region's population.
2. No single development type should be recommended for Amazonia, but rather a patchwork of different types, including diverse agro ecosystems and reserves of natural ecosystems.
3. All development alternatives have drawbacks. No development strategy provides a panacea, be it a single development type or a combination of types.
4. Development alternatives can be divided into

extensive and intensive uses, and into uses which maintain rainforest canopy cover and those which do not. Intensive use of cleared areas, preferably areas where forest cover has already been lost, has a number of advantages over extensive uses. Perennial crops providing tree cover, not all do, have better prospects than annuals. Markets and soils, among other limitations, greatly restrict the areas to which this form of management can be applied. Forest management schemes based on natural regeneration have the best prospects for producing sustainable economic returns in the vast areas of the region not suited to intensive farming. Pasture has the worst long-term prospects of all.

5 Social forces pressing in the direction of unwise land uses must be minimized by careful placement of areas zoned for different uses to reduce invasion risk for reserves or areas in non-intensive uses. Plans should be designed to require a minimum of enforcement, but necessary steps must be taken to assure compliance with needed regulations for protection of reserves and realization of long-cycle management schemes.

6 Long term well-being in the region requires that interdependent problems be solved simultaneously with the adoption of appropriate land uses. The social changes needed to reduce inequalities in income distribution and land tenure, and to maintain population levels below carrying capacity, must be brought about.

Activities linked to this chapter

Your teacher has details for the activities listed below:

* 'Evaluating development options in Amazonia' (Activity 13).
* 'Development Area role play' (Activity 14).

'Brazil is ours' explains the caption on this Yanomami girl's tee shirt. But, for her people, 'national integration' and 'national development', has meant death and cultural ruin. *(Claude Dumenil/Survival International)*

A view point from Marcus Colchester

Lessons to be learned and a strategy for Amazonia's survival

For the Indians of Amazonia the main lessons to be learned from contact with the outside world are simple. To survive they must retain, or regain, control over their traditional lands. Only once they have re-established their rights to these lands can they interact with the encroaching national societies from a position of security. Only with their lands do they retain choice, to choose which aspects of national and foreign culture and technology to incorporate into their own lives and which to reject.

For the outsiders the lessons are more difficult. The fragility of the tropical forests and the rate and consequences of their elimination have made westerners painfully aware of the destructiveness of their own society, of its unsustainable nature and ultimately of its end. The prospect of the death of the forest is a frightening indication of our own civilisation's mortality.

The debacle of Amazonian "development" has shown western society that technologically advanced systems are not for that reason superior or more desirable. The subtle balance between society and environment that the Indians have been able to maintain over millenia is founded on the achievement of an equilibrium between social, political, economic and technological forces. Change of any one aspect of this delicate balance can cause its disruption, which is not to say destruction of the entire system.

Our own society lacks any kind of balance or equilibrium. Change is the only constant of our time. Self-reliance on our own lands is no longer possible for us, nor, in spite of our aspirations to "democracy", can we hope to re-establish the kinds of egalitarian and fully representative political processes that are integral to Amazonian Indian life. But the lessons to be learned from Indian society are nevertheless relevant to us. Just as an understanding of the manner in which Indian society interacts with its environment depends on us understanding its social and political order as well as the strictly economic relations between society and environment, so is the same true for our own society.

The logging, cattle-ranching, mining, road building and colonisation programmes that are destroying Amazonia today are expressions of social and political forces far outside the region. Tropical hardwoods are cut for making plywood and veneers on colour televisions sold in Japan and Europe. Cattle are butchered in Amazonia's fast degrading grasslands to provide beef for our hamburgers. Cheap iron ore extracted with funds from the EEC is produced to maintain the competitiveness of the European steel industry. Roads built with money from the World Bank, which came from tax-payers in the "developed" world, provide the arteries for resettling thousands of pioneer farmers in Amazonia, whose agricultural systems are entirely inappropriate and unsustainable.

These people in turn have been made landless by vast agribusinesses, growing soya for export to service the country's debts to the First World, on what were once the small farmers' lands in the south of Brazil.

The pressure to open up and thus destroy the world's last wilderness regions can only be reduced by a radical change in our own systems of production and consumption. This is only likely to come about when the political relations between rich and poor, between north and south, undergo a major transformation.

The Indians are not passive victims of progress. Nationally and internationally they have mobilised to defend their lands and livelihoods. Here, Amazon Indian leaders Evaristo Nugkuag from Peru (right), and Cristobal Tapuy from Ecuador (left), with Survival International President, Robin Hanbury-Tenison, protest to the Museum of Mankind that the Indians are displayed as folklore rather as viable societies struggling to determine their own future. *(Sergio Dorantes/Survival International)*

REFERENCES

Allegretti, Mary Helena, and Schwartzman, Stephan, 'Extractive reserves: a sustainable development alternative for Amazonia', Report to WWF-US Project US – 478 (1986).

Barrow, C. 'The development of the varzeas of Brazilian Amazonia', In: *Man's impact on forests and rivers*, editor Hemming, J. (Manchester University Press, 1985).

Botafogo, J. Personal letter, *The Ecologist*, 15 (1985), 207–210.

Branford, S., and Glock, O., *The Last Frontier: fighting over land in the Amazon* (Zed, 1985).

Fearnside, P., 'Spatial deforestation in the Brazilian Amazon' *Ambio*, 15 (1986), 74–81.

Fearnside, P., 'Agriculture in Amazonia', In: *Amazonia*, editors Prance, G. and Lovejoy, T. (Pergamon Press, 1985).

Fearnside, P., 'Development alternatives in the Brazilian Amazon: an ecological evaluation', *Interciencia*, 8 (1983), 65–77.

Goodland, R., 'Brazil's environmental progress in Amazonian development', In: *Man's impact on forests and rivers*, editor Hemming, J. (Manchester University Press, 1985).

IUCN, UNEP and WWF, *World Conservation Strategy* (IUCN, 1980).

Johnson, B., *Towards a conservation and development strategy for the Amazon* (Earthlife, 1985).

Lanly, J.P., *UNEP/FAO Tropical Forest Assessment – a summary paper* (FAO, 1983).

Lutzenberger, J. 'The World Bank's Polonoroeste Project: a social and environmental catastrophe', *The Ecologist* 15 (1985), 69–72.

ACRONYMS

INCRA   Instituto Nacional de Colonizacao e Reforma Agraria (National Institute for Colonization and Agrarian Reform, Government of Brazil)

INPA   Instituto Nacional de Pesquisas da Amazonia (National Institute for Amazonian Research, Government of Brazil)

IUCN   International Union for Conservation of Nature and Natural Resources

MDB   Multi-lateral Development Bank

NGO   Non-Governmental Organisation

UNEP   United Nations Environment Programme

WCS   World Conservation Strategy

GLOSSARY

agouti   a rabbit-sized rodent.

agroforestry   a word used to describe the cultivation of trees on farms rather than in forests.

annual   a plant that completes its life cycle, from seed germination to seed production, followed by death, within a single season.

aroid   a species of plant from the family Araceae.

capybara   largest living rodent, live in groups near water.

cassava   also known as manioc. The floury pulp is baked and made into coarse bread.

peccary   a pig-like animal.

perennial   a plant that continues its growth from year to year (compare with annual)

sustainable development   economic development which takes account of the scarcity of resources with the intention of conserving or renewing them. This perspective argues that economic development in the longer term, if it ignores the *ecological* dimension, will undermine itself and will not improve a country's economic and social welfare.

terra firme   unflooded, upland rainforests of Amazonia.

varzea   floodplains of Amazonia.

ACKNOWLEDGEMENTS

We would particularly like to thank Christopher Joseph, Head of Geography, Marlborough College for editing the draft manuscripts and Adrian Cowell, Director of Central TV's "Decade of Destruction" series, for his help in supplying materials and photographs for this project.

Our thanks to the teachers involved with the WJEC A' level Geography syllabus who made valuable suggestions concerning the framework and contents of this publication:

Edryd Evans, Lewis Girls School, Hengoed
Cathie Roberts, Neath College, Neath
Norman Evans, Heolddu Comprehensive School, Bargoed
Joanne Thomas, Archbishop McGraph Comprehensive School, Bridgend
Alan Rosney, Coed-y-Ian Comprehensive School, Pontypridd
Jerry Foley, Bryntirion Comprehensive School, Bridgend
Anthony Preece, Bedwas Comprehensive School, Bedwas
David Smith, Cynffig Comprehensive School, Bridgend
Jac Thomas, St Ilan Comprehensive School, Caerphilly
Chris Davey, Tonyrefail Comprehensive School, Porth

The authors would also like to thank the following for their assistance during the development of "Rainforests":

John Huckle (Bedford College of Higher Education), Graham Hawkins (Humanities Advisor, Somerset LEA), Stephan Schwartzman, Jose Lutzenberger, Tony Dale (Oxford University Press), Tony Long (WWF), Philip Fearnside (INPA), J.P. Lanly (FAO) and staff at the World Bank's Office of Environmental and Scientific Affairs.

The field research on which the chapter on Indian land use is based was funded by The Social Science Research Council, the Ruggles – Gate Trust and the Emslie – Hornniman Foundation and carried out with the assistance of the Fundacion la Salle, Caracas and the Direccion de Asuntos Indigenas, Venezuela. This chapter also draws on the results of other anthropologists' research among the Yanomami.

Many thanks to Cherry Duggan and Wendy Hardy for word-processing the final documents and the numerous drafts that preceded them.

WWF is an international conservation foundation with national groups around the world, including Britain. It was launched in 1961 to raise money for the conservation of nature and natural resources. So far it has channelled over one hundred million pounds into projects in 135 different countries.

Through the funding of national parks, nature reserves, and environmental education provision, and through the promotion of conservation legislation and environmentally sensitive agricultural and industrial practices, WWF has protected individual species of animals and plants, as well as entire habitats such as rain forests, marshes, islands, meadow-land and coastal areas throughout the world.

WWF United Kingdom is committed to an education programme that aims to produce resource materials that enable teachers to bring environmental issues into everday classroom teaching at all levels of education. The materials are designed to give young people knowledge and experience that allows them to make informed personal judgements about environmental issues. Resources are being developed for most subjects of the school curriculum, making use of the inherent qualities of each subject to develop specific aspects of environmental understanding and sensitivity. These resources are produced from direct classroom work, from the work of groups of teachers, or by subject experts in conjunction with classroom trials.

If you would like full details about the education programme please send for the WWF United Kingdom "Materials for Teachers" catalogue (three 2nd class stamps essential).

> WWF – United Kingdom
> Education Dept
> Weyside Park
> Godalming
> Surrey GU7 1XR

Survival International is a world-wide movement that defends the rights of threatened tribal peoples to survival, self-determination and the use and ownership of their traditional lands.

Its network of members and supporters extends into 63 countries. Survival International campaigns for changes that are urgently needed; changes in attitudes, development projects, government policies and laws, in order to bring an end to violence, injustice and discrimination against tribal peoples.

> Survival International
> 310 Edgware Road
> London W2 1DY
> 01-723-5535